# Textile Science and Clothing Technology

**Series editor**

Subramanian Senthilkannan Muthu, SGS Hong Kong Limited, Hong Kong, Hong Kong

More information about this series at http://www.springer.com/series/13111

Subramanian Senthilkannan Muthu
Editor

# Sustainability in the Textile Industry

 Springer

*Editor*
Subramanian Senthilkannan Muthu
SGS Hong Kong Limited
Hong Kong
Hong Kong

ISSN 2197-9863         ISSN 2197-9871   (electronic)
Textile Science and Clothing Technology
ISBN 978-981-10-9674-7      ISBN 978-981-10-2639-3   (eBook)
DOI 10.1007/978-981-10-2639-3

Printed on acid-free paper

This Springer imprint is published by Springer Nature
The registered company is Springer Nature Singapore Pte Ltd.
The registered company address is: 152 Beach Road, #22-06/08 Gateway East, Singapore 189721, Singapore

# Contents

# Introduction

**Subramanian Senthilkannan Muthu**

**Abstract** Textile industry is one of the basic industries which satisfies one of the basic needs of people and hence becomes as an inevitable part of human's life. This book is planning to cover the detailed aspects of sustainability in textile industry encompassing environmental, social and economic sustainability in textiles and clothing sector. Beginning with the introduction to various faces and facets of sustainability, this book revolves around their implications to textiles and clothing sector. It is important and need of the hour to talk about the environmental impacts of textiles and clothing. But when it comes to sustainability of textiles and clothing, it is most commonly noticed from various sources such as literature and media that the other two pillars of sustainability namely social and economic are masked and carried away by the environmental impacts. Only environmental impacts are in focus when sustainability in textiles is defined. The other two parts or elements of sustainability are also equally important and they make much more sense when it comes to sustainability in textile industry. These uncharted areas will be discussed in detail to augment to the sustainability in textiles knowledge further.

**Keywords** Environmental sustainability · Economic sustainability · Social sustainability · Raw material · Manufacturing · End of life

## 1 Introduction

Sustainability in textile industry is a vast topic and it cannot be explained in a single volume of book, as this topic encompasses a wide array of elements into it. This topic is mainly chosen as a theme or title of this book for manifold reasons and one of them is that this title is becoming more and more important these days and many universities in the world have started teaching subjects in this area and with a hope that this book can become as a useful reference for the students. Second reason is

S.S. Muthu (✉)
Environmental Services Manager-Asia, SGS (HK) Limited, Hong Kong, China
e-mail: drsskannanmuthu@gmail.com

© Springer Nature Singapore Pte Ltd. 2017
S.S. Muthu (ed.), *Sustainability in the Textile Industry*,
Textile Science and Clothing Technology, DOI 10.1007/978-981-10-2639-3_1

that this topic is so diverse and there is a lot more to write. There are many aspects of this topic, which were rarely touched or there are no books on those aspects.

The concept of sustainability is not new and it has been known to the world since 1962 after the publication of Rachel Carson's Silent Spring. Publication of Silent Spring is the beginning of our thought process towards sustainability and especially the interactions among the basic elements of sustainability (SD Timeline 2012). Later on, there were many developments in the field of sustainability (Mebratu 1998) since today and the developments are going on and on every day. There were many activities which have fostered the concept and understanding of sustainability after the release of Silent Spring in 1962, such as the initiation of first Earth Day in 1970 (USA), polluter pays principle in 1971, release of work on chlorofluorocarbons (CFCs) by Rowland and Molina in Nature journal in 1974, release of World Conservation Strategy in 1980, establishment of World Resources Institute in 1982, discovery of Antarctic ozone hole in 1985. This long list also has the release of Brundtland Report—Our Common Future in 1987; this is the report of the World Commission on Environment and Development, which has united social, economic, cultural and environmental issues and global solutions together. Very important aspect of this report is that this is the one which has familiarized the term 'sustainable development'. Post-Brundtland report release, there were many important activities to enhance the sustainability thinking such as establishment of Intergovernmental Panel on Climate Change (IPCC) in 1988, Earth Summit—UN Conference on Environment and Development (UNCED) held in 1992, First meeting of the UN Commission on Sustainable Development in 1993, adoption of ISO 14001 as a voluntary international standard for corporate environmental management in 1996, release of reporting guidelines by Global Reporting Initiative in 2002, entering of Kyoto Protocol into force in 2005, development of Montreal Protocol on Substances that Deplete the Ozone Layer in 2007 (SD Timeline 2012).

What does sustainability really mean?—This is a million dollar question and one cannot get an unanimous answer from everyone if the question is asked among a wide array of people, and the debate on the definition of sustainability still continues. This is due to the matter of fact that sustainability and sustainable development (predecessor of sustainability) mean different to different people and the challenge of defining sustainability has already been noted by many authors already (White 2013). A well-known and very frequently read definition is the 'development which meets the needs of the present without compromising the ability of future generations to meet their own needs'. This was defined in the World Commission on Environment and Development's report—Our Common Future (World Commission on Environment and Development 1987). This definition was criticized heavily since its inception, as it is vague and open to interpretation. After this, there are a plenty of definitions one can see for sustainability and sustainable development and they were discussed elsewhere. One can see enormous amount of information from various media (Web, books, reports and research papers) on this.

In the quest of defining sustainability in simple terms, I read many books and papers in the literature for a considerable amount of time. Through my earnest

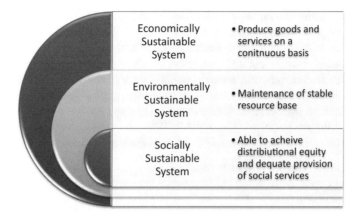

**Fig. 1** Sustainable system (*Source* Designed by Author from the reference Harris 2000)

efforts, I found a lot many and the very specific one which I would like to point out here is as follows:

> If reliable rainfall adds 100 l of rainwater to a tank every day, it is sustainable to use up to 100 l of this water per day. If the tank is large and is full to begin with, for a while it may be possible to use considerably more than 100 l per day. However, if the daily input from rainfall remains only 100 l, even starting with a full 10,000-l tank, one cannot use more than 100 l per day sustainably. The tank will eventually run dry. (Population Matters 2011)

Concentration of the crux of sustainability varies between different professionals when they define sustainability. Environmentalists and ecological professionals define sustainability with a chief focus on environmental side, and businessmen shift the focus on economic side and so on. However, as said earlier, it has three dimensions–economic, social and environmental, and all the three have to be fulfilled (Rankin 2014). Figure 1 is the reflection of a sustainable system fulfilling all these three pillars into the core concept of sustainability (Harris 2000).

It is a great challenge to manage all the three issues together for today's managers (CSR managers). It is highly important and need of the hour for companies to understand that they are not detached, rather they are part of a actuality that demands the management of scarce resources and as well as the concern for social issues (Savitz and Weber 2006).

## 2  Environmental Sustainability

Environmental sustainability can simply be defined as 'the ability to maintain things or qualities that are valued in the physical environment', where the physical environment includes the natural and biological environments (Sutton 2004). 'Environmental sustainability' requires maintaining natural capital as both a provider of economic inputs called 'sources' and an absorber called 'sinks' of economic outputs called 'wastes' (Daly 1973, 1974; World Bank 1986; Serageldin 1993; Pearce and Redclift 1988; Pearce et al. 1990a, b).

Environmental sustainability refers to ecosystem integrity, carrying capacity and biodiversity. It demands that natural capital be preserved to be a foundation of economic inputs and as a sink for wastes. Resources need to be harvested no faster than they can be regenerated. Wastes must be emitted no faster than they can be digested by the environment (Kahn 1995). When it comes to embedding environmental sustainability into a sustainable system, such system must be able:

- To maintain a stable resource base;
- To avoid over-exploitation of renewable resource systems (or environmental sink functions);
- To evade the depletion of non-renewable resources to the possible extent;
- To deplete non-renewable resources only to the extent that investment is made in adequate substitutes (necessarily includes maintenance of biodiversity, atmospheric stability and other ecosystem functions not ordinarily classed as economic resources) (Harris 2003).

## 3 Economic Sustainability

'Economic sustainability' refers to a system of production that satisfies present consumption levels without compromising future needs (Kahn 1995). On a system level, an economically sustainable system must be able to produce goods and services on a continuing basis, to maintain manageable levels of government and external debt and to avoid extreme sectoral imbalances which damage agricultural or industrial production (Harris 2003).

## 4 Social Sustainability

Looking at a basic sense, 'social sustainability' implies a system of social organization that assuages poverty. In a deeper fundamental sense, however, 'social sustainability' establishes the nexus between social conditions such as poverty and environmental decay (Ruttan 1991; Basiago 1999). When it comes to system level, a socially sustainable system must achieve fairness in distribution and opportunity, adequate provision of social services including health and education, gender equity, and political accountability and participation (Harris 2003).

# 5 Sustainability in Textile Supply Chain

Textile industry consists of a highly complex and a massive supply chain (Basiago 1999; Muthu 2014a, b; Gardetti and Muthu 2015). Diversity and the intermittent supply chain add further severity to it. There are many processes involved in getting an apparel product finally manufactured and there are many partners playing a key role in this lengthy supply chain to produce a clothing product. This industry is such a diversified one with many fibres and their dedicated process lines, and even with the same process line (say for cotton sector manufacturing line or woollen line), there are many variants in the processes. A very detailed process chain for various fibres, spinning methods, fabrication methods, chemical processing and apparel manufacturing steps is already explained in my previous publications (Sutton 2004). A very simple flow chart, which shows the key operations and players in textiles and clothing supply chain, is depicted in Fig. 2. Inputs marked in Fig. 2 as number 1 refer to raw materials, energy, water, chemicals, auxiliaries and even human inputs (labour). Outputs marked in Fig. 2 as number 2 refer to finished products, emissions (to air, water and land), wastewater and solid waste.

As indicated in Fig. 2, finished products reach consumer after sales efforts and after the consumer uses and decides to dispose it might join the same supply chain back as inputs if the product is recycled (as a closed loop). This again touches the life cycle of textile products, which will be discussed in detail in the forth coming chapters; however, at this stage the discussions are pertinent to the entire supply chain. Life cycle of textile products includes the following phases:

- Fibre cultivation or manufacturing;
- Textile and clothing production;
- Retailing;
- Consumer use;
- End of life.

Sustainability aspects of textiles must look into the entire supply chain of textiles and the clothing with the scales to assess and further to improve upon the social,

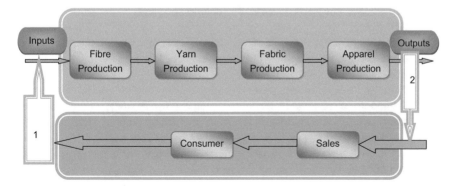

**Fig. 2** Textiles and clothing supply chain

environmental and economic impacts at each stage of the entire supply chain. Talking about environmental impacts particularly, there is a long list of environmental impacts which can be evaluated for various textile products in their different life cycle phases. Some of these impacts include, but not limited to:

- Climate change or carbon footprint;
- Ecological footprint;
- Acidification;
- Eutrophication;
- Human toxicity;
- Ecotoxicity;
- Water depletion;
- Energy demand;
- Depletion of abiotic and biotic resources;
- Ozone depletion potential;
- Photochemical smog;
- Land use impacts.

Textiles and clothing sector is highly entangled with environmental, social, economic and also governmental issues. Producers and retailers have been focusing their efforts in reducing the environmental impacts of textiles in their entire life cycle stages and also improving the social aspects (such as instituting fair working conditions, setting social standards, establishing minimum wages, ensuring occupational safety, imposing a ban on child and forced labour and so on) (Sustainability of Textiles 2013).

There are umpteen number of challenges textiles and clothing industry is facing as far as sustainability in the entire supply chain is considered. Some of the main challenges are as follows (Sustainability of Textiles 2013; Diviney and Lillywhite 2009):

- Environmental impacts—this varies from fibre to fibre. However, the main environmental impacts challenging the current textile industry are as follows:
  - Significant energy use in the entire manufacturing link;
  - Greenhouse gas (GHG) emissions (carbon footprint) in the entire manufacturing link;
  - Significant water use in fibre manufacturing and production stages (fibre-specific issue will be discussed in detail later);
  - Ecotoxicity from washing and drying of textiles;
  - Toxicity from fertilizers, pesticides in the fibre stage of natural textiles;
  - Depletion of renewable resources such as fossil fuels, energy use and associated GHG emissions in the fibre production stage of synthetic fibres;
  - Release of nutrients (which is responsible for eutrophication);
  - Toxicity (human and ecotoxicity), hazardous waste substances management and effluent treatment during the production stage and with the employment of chemicals, dyes and finishes in the manufacturing link;

- Cost of sustainable production;
- Fast fashion cycles;
- Consumer behaviour;
- Social criteria including working conditions, child labour, poor wages, safety and so on;
- Environmental, health and safety issues;
- Textile waste management coupled with landfill shortage;
- Non-degradable textile materials;
- Economic issues in the entire supply chain and also in the trade.

The above list is not exhaustive and there are many other elements too to join this list. One can imagine that for such a long list a quick-fix and ready-hand solution is not possible and adding to it, there are multiple players involved in this entire loop to decide on various issues leading to sustainability.

## Recommendations for Further Study

Ayres RU, Ayres LW (1996) Industrial ecology: closing the materials cycle. Edward Elgar, Cheltenham

Basiago AD (1995) Methods of defining 'sustainability'. Sustain Dev 3(3):109–119

Beckerman W (1994) Sustainable development: is it a useful concept? Environ Values 3:191–209

Carpenter RA (1995) Limitation in measuring ecosystem sustainability. In: A sustainable world: defining and measuring sustainable development, T.C. Tryzna. Earthscan Publications, London

Clarke JJ (1993) Nature in question. Earthscan Publications, London

Clayton AMH, Radcliffe NJ (1996) Sustainability—a systems approach. Earthscan Publications Ltd.

Cooper DE, Palmer JA (eds) (1992) The environment in question: ethics and global issues. Routledge, London

Daly HE (1990a). Boundless bull. Gannett Cent J 4(3):113–118; Daly HE (1990b) Toward some operational principles of sustainable development. Ecol Econ 2:1–6

Kirkby J, O'keef P, Timberlake L (1995) Sustainable development: the earthscan reader. Earthscan Publications, London

Schvaneveldt SJ (2003) Environmental performance of products: benchmarks and tools for measuring improvement. Benchmarking—Intl J 10(2):136–151

## References

Basiago AD (1999) Economic, social, and environmental sustainability in development theory and urban planning practice. The Environmentalist 19:145–161

Daly HE (ed) (1973) Towards a steady state economy. Freeman, San Francisco

Daly HE (ed) (1974) The economics of the steady state. American Economic Review, 15–21 May

Diviney E, Lillywhite S (2009) Travelling textiles, a sustainability roadmap of natural fibre garments, Brotherhood of St Laurence, Australia, May 2009

Gardetti MA, Muthu SS (2015) Sustainable apparel? Is the innovation in the business model? The case of IOU project. Text Cloth Sustain 1:2

Harris JM (2000) Basic principles of sustainable development, global development and environment institute. In: Working Paper 00-04, June 2000, Tufts University Medford, Medford

Harris JM (2003) Sustainability and sustainable development, international society for ecological economics, internet encyclopedia of ecological economics, Feb 2013, Available at: http://isecoeco.org/pdf/susdev.pdf. Accessed 10 June 2015

Kahn M (1995) Concepts, definitions, and key issues in sustainable development: the outlook for the future. In: Proceedings of the 1995 international sustainable development research conference, Manchester, England, Mar 27–28 1995, Keynote Paper, 2–13

Mebratu D (1998) Sustainability and sustainable development: historical and conceptual review. Environ Impact Asses Rev 18:493–520

Muthu SS (2014a) Assessing the environmental impacts of textiles and the clothing supply chain. Woodhead Publishing, UK

Muthu SS (2014b) Roadmap to sustainable textiles & clothing, environmental and social aspects of textiles and clothing supply chain, Preface, V

Pearce DW, Redclift M (eds) (1988) Sustainable development. Futures 20 (Special Issue)

Pearce DW, Barbier E, Markandya A (1990a) Sustainable development: economics and environment in the third world. Elgar, Aldershot

Pearce DW, Markyanda A, Barbier A (1990b) Blueprint for a green economy. Earthscan, London

Population Matters (2011) Sustainability and the Ehrlich equation, glossary indepth, 2011. Available at: https://www.populationmatters.org/wp-content/uploads/ipat.pdf, Accessed 9 June 2015

Rankin WJ (2014) Chapter 4.1: Sustainability, treatise on process metallurgy, vol 3, pp 1376–1424

Ruttan VW (1991) Sustainable growth in agricultural production: poverty, policy and science. Unpublished paper prepared for International Food Policy Research Institute Seminar on Agricultural Sustainability, Growth, and Poverty Alleviation, Feldafing, Germany, Sept 23–27

Savitz AW, Weber K (2006) The triple bottom line: how today's best-run companies are achieving economic, social, and environmental success—and how you can too. Wiley, New York

SD Timeline (2012) The International Institute for Sustainable Development

Serageldin I (1993) Developmental partners: aid and cooperation in the 1990s. SIDA, Stockholm

Sustainability of Textiles (August 2013) ISSUE PAPER N° 11, Retail forum for sustainability. Accessed from: http://ec.europa.eu/environment/industry/retail/pdf/issue_paper_textiles.pdf

Sutton P (2004) A Perspective on environmental sustainability? A paper for the victorian commissioner for environmental sustainability, green innovations, Australia, April 2004

White MA (2013) Sustainability: i know it when I see it, commentary. Sustainable urbanization: a resilient future. Ecol Econ 86:213–217

World Bank (1986) Environmental aspects of Bank Work. The World Bank operations manual statements, OMS 2.36. World Bank, Washington, DC

World Commission on Environment and Development (1987) World commission on environment and development-our common future. Oxford University Press, New York

# Evaluation of Sustainability in Textile Industry

## Subramanian Senthilkannan Muthu

**Abstract** Sustainability and its assessment are highly critical to industries, governments and even to customers. Every industrial sector, company, individual and even government and different nations have sustainability goals and commitments. It is inevitable to measure the level of sustainability achievement by all these above-mentioned players and for this, a tool to assess sustainability is highly desirable. Assessment of sustainability is an important topic in any field. Choosing a right tool to suit the needs of different players in the field of sustainability is very much crucial. As mentioned in earlier chapter, sustainability focuses on three major elements namely environmental, economic and social aspects and any assessment technique needs to focus on this triple-line thinking. Sustainability can be assessed for a product/process, project and also for a sector and a country. Assessment tools are classified by various ways. Various authors have classified these tools in various ways according to the ruler/scale/method. Not all tools can be used for textiles and clothing supply chain, and some of the tools are highly evident to assess the sustainability of textile products. This chapter deals with various sustainability tools and their implications in textile industry.

**Keywords** Tools · Product assessment · Life cycle assessment · Life cycle costing · Social life cycle assessment · Textile sector

# 1 Classifications of Assessment Tools

According to one of the methods of classification, there are three types of assessment tools existing namely product-related assessment, project-related assessment and sector- and country-related assessment. Additionally, indicators or indices also join this list (Rorarius 2007; Štreimikienė et al. 2009). Other classification method

S.S. Muthu (✉)
SGS Hong Kong Limited, Tsuen Wan, Hong Kong
e-mail: drsskannanmuthu@gmail.com

© Springer Nature Singapore Pte Ltd. 2017
S.S. Muthu (ed.), *Sustainability in the Textile Industry*,
Textile Science and Clothing Technology, DOI 10.1007/978-981-10-2639-3_2

classifies the tools again into three types namely monetary, biophysical and indicator-based tools (Gasparatosa and Scolobig 2012). The third type classifies also into three types namely indicators and indices, which are further classified into non-integrated and integrated, product-related assessment tools with the major focus on the material and/or energy flows of a product or service from a life cycle perspective, and finally integrated assessment (Bebbington et al. 2007; Ness et al. 2007).

Back to the basics, any assessment tool for sustainability must focus on three elements namely environmental sustainability, economic sustainability and social sustainability. There are many dedicated tools, and methods were developed for assessing each element of sustainability. One of the major areas which needs to be discussed when it comes to sustainability assessment is life cycle assessment or life cycle thinking. This will be discussed chiefly in this section.

In the above classification of tools, one can understand that the indicators are one of the essential tools captured by all the 3 classification methods mentioned in this book in this section (please note that these three types mentioned here are not exhaustive; there might be many types of classifications in addition to the three types pointed out here). There are many indicators and indices defined by several researchers in the field of sustainability, and one can imagine a list as high as 140 indicators defined by UN Commission on Sustainable Development (CSD) (CSD 2001).

## 2 Life Cycle Assessment (LCA) and Life Cycle Thinking (LCT)

Life cycle assessment (LCA) is a scientific method or technique to identify, quantify and further to evaluate the environmental impacts (inputs and outputs) of a product, service or activity, from cradle to grave stages (Life Cycle Analysis 2015). It can be used to quantify the environmental impact of processes as well and again it has many variants. As the name implies, it is highly helpful to assess the environmental impacts of a product from the beginning (raw material extraction stage or cradle) to the final (disposal or end of life or it is called as grave stage). Life cycle of a typical product comprises the five stages as depicted in Fig. 1.

Not always a LCA study has to include all the stages from cradle to grave. At times, the depth of the study can be shortened and confined from cradle to gate or gate to gate stages. Likewise, there are many such variants of LCA such as well to wheel. A very detailed explanation of LCA and how a LCA study needs to be conducted and what are the challenges and limitations pertaining to LCA studies in textiles are already explained in depth by the author in his other books (Life Cycle Analysis 2015; Muthu 2014; Muthu 2015a). Instead of repeating the same here, very important points are only mentioned in this section. Readers are strongly encouraged to read those books to get detailed information on LCA studies on textiles and clothing sector.

**Fig. 1** Life cycle stages of a
typical product

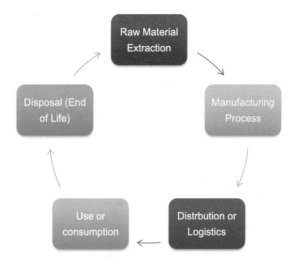

Life cycle assessment is a multi-step process to calculate the environmental impacts (potential) created by a product in its lifetime. ISO standards have a dedicated series for LCA methodology, and the important widely used and accepted standards for LCA are ISO 14040 and 14044. There are many other variants of LCA such as product carbon footprint assessment (assessed by ISO/TS 14067), which is again based on LCA technique and relying on ISO 14040/44. According to ISO series of LCA standards, a LCA study necessarily has four steps (Fig. 2).

**Fig. 2** Four-step process
of LCA

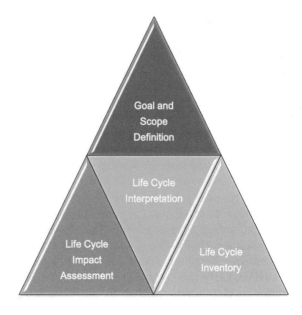

This concept of LCA can be used to measure different types of sustainability, and there are many tools developed in line with life cycle thinking to measure different dimensions (economic, social and environmental) sustainability. Life cycle thinking (LCT) is the concept of including the whole product life cycle system from the "cradle to the grave" under the consideration (raw material extraction stage to the end of life stages) with the objective of preventing individual life cycle stages are considered for assessment which will subsequently shift the environmental burdens to another life cycle stage (Finkbeiner et al. 2010).

## 3 Life Cycle Approach—Sustainability Assessment

As discussed in the previous section, to measure the three elements or faces of sustainability namely environmental, economic and social, life cycle approach is widely used. There are different types of tools developed to measure these three sustainability elements, and based on the element of sustainability to be assessed, an appropriate tool has to be chosen.

To assess the environmental impacts of a product, LCA is widely used. LCA typically refers to environmental LCA. As discussed above, LCA follows cradle to grave concept with cross-media environmental assessment approach, i.e. all relevant environmental impacts such as both on the input side (consumption of resources) and on the output side (emissions to air, water and soil, including waste) (Finkbeiner et al. 2010). If the life cycle stages of a product are assessed with an environmental lens, then it is referred to as environmental LCA. LCA is defined by ISO 14040 as "compilation and evaluation of the inputs, outputs and the potential environmental impacts of a product system throughout its life cycle" (ISO 14040 2006).

Life cycle costing refers to the assessment technique if different life cycle phases of a product from its cradle to grave stages are examined with the aid of an economic lens. Life cycle costing is defined as summing up total costs of a product, process or activity discounted over its lifetime (Spitzer and Elwood 1995; Henn 1993; Spitzer et al. 1993; EPA 1993). For economic sustainability assessment, two factors are mainly considered—cost and performance—and there are variety of approaches and tools developed for the same. Manufacturing costs (from a business perspective) and life cycle costs (from the customer's perspective) are mainly considered for economic evaluation (Finkbeiner et al. 2010; Publications of the European Platform of LCA 2010).

Life cycle costing (LCC) is a methodology that enables you to incorporate costs and benefits that occur over the entire life cycle of a product into your procurement decisions, rather than considering the initial capital cost of a product only (Mungcharoen 2013). The process of life cycle costing (LCC) fundamentally involves investigating the costs arising from an asset over its entire life cycle and also investigating and suggesting the alternatives that have an impact on this cost of ownership (Life Cycle Costing 2011). Australian Standard AS/NZS 4536:1999 defines life cycle cost as the sum of acquisition cost and ownership cost of an asset over its life cycle from design stage, manufacturing, usage, maintenance and disposal (Australian Standard AS/NZS 1999).

Life cycle costing helps to ascertain the costs of any product in different life cycle stages, and it offers great help in green public procurement (GPP). It is a misconception that green or sustainable products will cost more and if we have a deeper dive into this, this does not necessarily be true. Even though in many cases (of course not all the time), the sustainable alternatives or green choices may have higher initial or purchase price. However, when we interrogate into the total cost throughout the entire life cycle of the product, these sustainable/green choices may prove to be cheaper than the nongreen alternatives. Green alternatives may compensate the higher initial cost in the further life cycle stages such as usage/waste/disposal stages when compared to their rivals (Life Cycle Costing 2008). This kind of knowledge or information can be disseminated through LCC, and it can be a wonderful tool in textile industry, where there a lot of sustainable/green alternatives are coming in every day. However, the applicability of LCC in textiles and clothing industry is still at infancy state.

The third face of sustainability, which is social face, can also be assessed with the aid of life cycle approach. Such approach/methodology is called as social life cycle assessment (S-LCA). Contemplation of integrating social aspects into the traditional life cycle assessment of products started in the beginning of 1990s itself. Social life cycle assessment (S-LCA) assesses the potential positive and negative social impacts along a product's life cycle, while avoiding shifting negative impacts from one part of the supply chain to another.

S-LCA method can be used to evaluate the social and sociological aspects of products, their actual and potential positive as well as negative impacts along the life cycle. This assessment encompasses various life cycle stages of a product beginning from the raw material extraction and processing and progresses through the subsequent phases namely manufacturing process, transportation and distribution, consumer use and end of life (reuse, recycling and disposal). S-LCA utilises generic and site-specific data, which can be quantitative, semiquantitative or qualitative. S-LCA complements the environmental LCA and LCC, and it can either be applied alone or in combination with the other techniques (S-LCA; UNEP).

S-LCA method is becoming popular recently, and the methodological part pertaining to every industry is also getting familiarised and many researchers are now working into this field. Readers are suggested to refer to the author's recent book on Social Life Cycle Assessment for further details (Muthu 2015b).

Textile industry is facing a huge challenge in terms of social impacts especially working conditions, child labour and so on. Though there are a very few (one or two only to the knowledge of the author even after the vast literature search) studies on textiles to measure the social impacts of clothing products, an earmarked tool or methodological choice for measuring the social impacts of textiles and clothing products is yet to be developed.

Integrating all the three tools for assessing the three faces of sustainability is attempted and coined under the umbrella of life cycle sustainability assessment (LCSA).

LCSA = (environmental) LCA + LCC + S-LCA (Klöpffer 2008; Finkbeiner et al. 2010)

Combining (environmental) LCA, S-LCA and LCC contributes to an assessment of products, providing more relevant results in the context of sustainability (UNEP/SETAC Life Cycle Initiative 2011). However, such tool is yet to be attempted for measuring the overall sustainability of textile products.

## Recommendations for Further Study

Atkinson GD, Dubourg R, Hamilton K, Munasignhe M, Pearce DW, Young C (1997) Measuring sustainable development: macroeconomics and the environment. Edward Elgar, Cheltenham

BogenstGatter U (2000) Prediction and optimisation of life-cycle costs in early design. Build Res Inf 28(5/6):376–386

Bohringer C, Jochem PEP (2007) Measuring the immeasurable—a survey of sustainability indices. Ecol Econ 63:1–8

Bossel H (1999) Indicators for sustainable development: theory, method, applications. A report to the Balaton Group. IISD, Canada

Dreyer L, Hauschild M, Schierbeck J (2006) A framework for social life cycle impact assessment. Int J Life Cycle Assess 11(2):88–97

Hermann BG, Kroeze C, Jawjit W (2007) Assessing environmental performance by combining life cycle assessment, multi-criteria analysis and environmental performance indicators. J Clean Prod 15:1787–1796

Jørgensen A (2013) Social LCA-a way ahead? Int J Life Cycle Assess 18:296–299

Norris GR (2006) Social impacts in product life cycles-towards life cycle attribute assessment. Int J Life Cycle Assess 11(1)(Special Issue):97–104

Warren JL, Weitz KA (1994) Development of an integrated life-cycle cost assessment model. IEEE, New York, pp 155–163

Weidema BP (2006) The integration of economic and social aspects in life cycle impact assessment. Int J Life Cycle Assess 11(1)(Special Issue):89–96 (CrossRef)

## References

Australian Standard AS/NZS 4536:1999, Life Cycle Costing: An Application Guide, Reconfirmed on 2014

Bebbington J, Brown J, Frame B (2007) Accounting technologies and sustainability assessment models. Ecol Econ 61:224–236

CSD (2001) Indicators of sustainable development: guidelines and methodologies. Commission on Sustainable Development, New York, USA. http://www.un.org/esa/sustdev/natlinfo/indicators/indisd/indisd-mg2001.pdf

EPA (1993) Life cycle design guidance manual: environmental requirements and product system. EPA-600-R-92-226, 1993, pp 122–129

Finkbeiner M, Schau EM, Lehmann A, Traverso M (2010) Towards life cycle sustainability assessment. Sustainability 2(10):3309–3322. Open access doi:10.3390/su2103309

Gasparatosa A, Scolobig A (2012) Choosing the most appropriate sustainability assessment tool. Ecol Econ 80:1–7

Henn CL (1993) The new economics of life cycle thinking. IEEE, New York

ISO 14040 (2006) Environmental management—life cycle assessment—principles and framework

Klöpffer W (2008) Life cycle sustainability assessment of products. Int J Life Cycle Assess 13 (2):89–95

Life Cycle Analysis, obtained from: http://www.gdrc.org/sustdev/concepts/17-lca.html. Accessed 10 Aug 2015

Life Cycle Costing, Better Practice Guide, Australian National Audit Office, December 2011, Obtained from: http://www.anao.gov.au/uploads/documents/Life_Cycle_Costing.pdf

Life Cycle Costing—Fact Sheet, European Commission Green Public Procurement (GPP) Training Toolkit—Module 1: Managing GPP Implementation, Toolkit developed for the European Commission by ICLEI—Local Governments for Sustainability, 2008, European Commission, DG Environment-G2, B-1049, Bruxelles

Mungcharoen T (2013) Approach on Life Cycle Costing and Benefits, Regional Workshop: Green Public Procurement and Eco-labeling, 1 May 2013 at Phuket, Thailand Organized by PCD, GIZ in collaboration with TEI, FTI, TGO & UNEP. http://www.thai-german-cooperation.info/download/201305_ecolabel_1530_thamrongrut_LCC.pdf

Muthu SS (2014) Assessing the environmental impact of textiles and the clothing supply chain, 1st edn. Woodhead Publishing (March 2014)

Muthu SS (2015a) Handbook of life cycle assessment (LCA) of textiles and clothing, 1st edn. Woodhead Publishing (July 2015)

Muthu SS (2015b) Social life cycle assessment—an insight. Springer Science+Business Media Singapore

Ness B, Urbel Piirsalu E, Anderberg S, Olsson L (2007) Categorising tools FOS sustainability assessment. Ecol Econ 60:498–508

Publications of the European Platform of LCA Including the ILCD Handbook; European Commission: Brussels, Belgium, 2010. http://lct.jrc.ec.europa.eu/publications

Rorarius J (2007) Existing assessment tools and indicators: building up sustainability assessment. Some perspectives and future applications for Finland. Finland's Ministry of Environment, Report

Social Life Cycle Assessment (S-LCA), Life Cycle Initiative, obtained from: http://www.lifecycleinitiative.org/starting-life-cycle-thinking/life-cycle-approaches/social-lca/

Spitzer M, Elwood H (1995) An introduction to environmental accounting as a business management tool: key concepts and terms. EPA, Washington, DC

Spitzer M, Pojasek R, Robertaccio F, Nelson J (1993) Accounting and capital budgeting for pollution prevention. United States Environmental Protection Agency. In: The engineering foundation conference, 24–29 Jan 1993, San Diego, CA

Štreimikienė D, Girdzijauskas S, Stoškus L (2009) Sustainability assessment methods and their application to harmonization of policies and sustainability monitoring. Environ Res Eng Manag 2(48):51–62

UNEP Guidelines for Social Life Cycle Assessment of Products. Accessed from http://www.unep.org/pdf/DTIE_PDFS/DTIx1164xPA-guidelines_sLCA.pdf

UNEP/SETAC Life Cycle Initiative (2011) Towards a Life Cycle Sustainability Assessment: Making informed choices on products. ISBN: 978-92-807-3175-0

# Environmental Sustainability in the Textile Industry

Kyung Eun Lee

**Abstract** The textile industry creates significant environment effects throughout the life cycle of textile products. This chapter explains strategic approaches for promoting environmentally sustainable textile product consumption and production in the textile industry. More specifically, five phases in environmental sustainability (material, manufacturing, retail, consumption, and disposal phase) are focused on in a discussion. Key concepts (e.g., corporate social responsibility, green supply chain management, and eco-design) for environmentally sustainable business practices to be considered are suggested in this chapter. It is important that all stakeholders, including consumers, manufacturers, supply chain, and retailers, in the textile industry take alternatives to promote environmental protection in the production and consumption of textile products.

**Keywords** Environmental sustainability · Green supply chain management · Corporate social responsibility · Eco-design

## 1 Introduction of Environmental Sustainability in the Textile Industry

In entrepreneurial points of view, environmental sustainability is a business strategy for using processes without generating harmful effects to the environment and natural resources throughout the life cycle (e.g., consisting of collection, processing, application, replenishment, consumption, and disposal) of natural resources (Roy Choudhury 2015; Khan and Islam 2015). For instance, a cotton T-shirt generates environmental impacts on these life cycle processes. In collection, cotton cultivation involves the use of fertilizer, pesticides, and water; in processing, to dye and spin cultivated cotton, coal, dyes, and auxiliaries, as well as water and

K.E. Lee (✉)
Department of Apparel, Events and Hospitality Management,
Iowa State University, 28 Mackay Hall, Ames, IA 50011, USA
e-mail: Kyungeun@iastate.edu

© Springer Nature Singapore Pte Ltd. 2017                                    17
S.S. Muthu (ed.), *Sustainability in the Textile Industry*,
Textile Science and Clothing Technology, DOI 10.1007/978-981-10-2639-3_3

electricity, are used; in application, T-shirt manufacturing requires the use of electricity, detergent, and water; in replenishment, to procure a reorder of T-shirts, collection, processing, and application are repeated; in consumption, end users purchase and wear T-shirt products; and in disposal, end users discard used T-shirts and manufacturers get rid of produced T-shirts or raw material left over (Zhang et al. 2015).

Hence, in the textile industry, the use of energy, chemicals, and water are major environmental impact generators throughout the life cycle of products. To ensure environmental sustainability, apparel designers should create products based on environmentally and socially responsible design approaches and trends; the supply chain must consider its impacts on society, economy, and environment for their business practices (Adams and Frost 2008; Roy Choudhury 2015).

There are some textile companies that have set environmental sustainability as a priority in their business practices. The Global 100 Most Sustainable Corporations in the World Index released by Corporate Knights, a media and research group in Toronto, consist of select companies in accordance with the scores obtained from four screening tests evaluating 12 key performance indicators of environmental sustainability (e.g., energy and water usage efficiency), as well as social responsibility (e.g., employee turnover and leadership diversity) (Kim 2016). The 2016 Global 100 included the following companies: in textile companies, Adidas (ranked two) and H&M (ranked five); in the beauty sector, L'Oréal (ranked 14) and Unilever (ranked 47); for retailers, Marks & Spencer (ranked 21); and in the luxury sector, the French luxury group Kering, which operates multiple European brands such as Gucci, Saint Laurent, and Balenciaga (ranked 43) (Kim 2016).

## 2 Environmental Impacts of the Textile Industry

The textile industry creates one of the most significant impacts to the global economy and the environment. Global textile industry's annual sales (including the apparel and footwear sectors) are expected to exceed US \$2 trillion by 2018 (Marketwired 2014). In 2015, the global textile industry made \$618 billion annual revenue with a 2 % growth rate and employed 5.8 million people (IBISworld 2015). Due to the significance of the market size, the textile industry is one of the greatest harmful effect generators to the environment with the use of chemical materials and processes (Gardetti and Muthu 2015). For instance, globally, 40 % of clothing is made with natural cotton fiber, one of the most chemically dependent crops consuming 10 % of all chemicals and 25 % of the insecticides used in worldwide agricultural industries (Sweeny 2015).

The accumulated harmful chemical residuals used in textile production are released, often untreated and directly into water sources, will eventually destroy the soil, water, and the environment (Oecotextiles N.D.). Together with the cultivation of raw materials, in the textile industry, the harmful effects are created throughout the life cycle of textile products including raw material production processes (e.g., fibers, yarns, and textiles), garment production processes (e.g., assembly and packaging), and consumption of manufactured textile products (e.g., end use, recycling, and discarding) (Khan and Islam 2015). For instance, in 2012, the denim industry occupied approximately 30 % in the total global apparel industry (Greenpeace 2012). Due to the excessive use of synthetic indigo dyeing methods, over 70 % of water was contaminated in the world's greatest denim-producing country, Mexico; hence, water scarcity has become a serious problem in Mexican society (Greenpeace 2012).

In addition to chemical discharge into water sources, during textile manufacturing processes, such as dyeing, printing, and finishing, an excessive amount of water, fossil fuels, and electrical energy is consumed (Oecotextiles 2013; Sivaramakrishnan 2009). Each year, over a half trillion gallons of freshwater is used in textile dyeing and approximately 70 million barrels of oil are consumed for virgin polyester production to be used in fabric manufacturing. As an example of the results of water consumption in the textile industry, the Aral Sea, the water source for the world's sixth largest cotton producer, Uzbekistan, currently has only 10 % of the water levels that were available 50 years ago (Sweeny 2015).

Fast fashion is a retailing approach capable of creating new trendy merchandise with quick turnover in retail stores (Loeb 2015). In 2013, the global fashion market size was 192,334 million euro with an 2.75 % annual growth rate that accounts for approximately 11 % of the total global apparel market (Fashionbi 2013; Marketwired 2014). Due to rapid product changeability and poor quality, sometimes fast fashion significantly contributed to creating a non-eco-friendly disposable clothing culture in which consumers dispose of clothing after wearing it once or twice; such a clothing culture impacts the packing of landfill spaces on the earth (Chau 2012).

# 3 Environmentally Sustainable Consumption and Production in the Textile Industry

As a result of severe environmental impacts created by the textile industry, environmental sustainability has become a fundamental concern for textile manufacturers' businesses and consumers' lifestyles and product purchase choices (Khan and Islam 2015). Hence, textile companies should develop initiatives to motivate their stakeholders (e.g., the owners, supply chains, and retailers) to participate in eco-friendly fashion business practices (Abreu 2015). There are three major forms of initiatives of textile companies for environmental sustainability: (a) corporate social responsibility (CSR), (b) green supply chain management (GSCM), and (c) eco-design.

## 3.1 Corporate Social Responsibility (CSR)

Corporate social responsibility (CSR) is a company's obligation to perform their business activities in a manner that is not harmful to the society or to the environment (Steiner and Steiner 2009). To execute CSR, firms assess their production in terms of economic, social, and environmental aspects (Taylor 2015). There are various forms of CSR such as profit sharing with nonprofit organizations, product giveaways for every sale made, and the use of sustainably grown raw materials or ingredients. For example, the company Ben and Jerry's only uses ingredients produced by their dairy farm sustainability program; Starbucks also sources coffee grown and processed based on their eco-friendly practice guidelines of C.A.F.E. program; Toms Shoes donates one pair of shoes to a child in need whenever a customer buys a pair (Taylor 2015).

From an environmental standpoint, a primary focus of CSR is protecting the environment, with a main focus on carbon footprint reduction. Currently, the textile industry consumes a quarter of the chemicals produced by all industries in the globe (Conca 2015). As a result, the textile industry creates over 10 % of global carbon emissions, and from cheap synthetic fiber productions, also emits toxic gases, such as N2O, 300 times more environmentally destructive than $CO_2$ (Conca 2015). To reduce these carbon emissions, CSR encourages textile companies to invest their efforts and budgets toward eco-friendly business practices.

Epstein-Reeves (2012) addressed multiple benefits of CSR for the firms: (a) innovation, (b) cost savings, (c) brand differentiation, and (d) long-term thinking. First, for innovation, companies' research and development efforts can create innovative products in which the core is environmental sustainability (e.g., Unilever's hair conditioner consuming less water). Second, CSR allows the firms to save costs through reduced packaging and energy usage (e.g., Levi's Water<Less™ jeans saving up to 92 % of water for production) (Levis N.D.). Third, incorporating sustainability into the company's business model can differentiate brands (e.g., Patagonia's 1 % for the planet program donates 1 % of sales for environmental protection) (Patagonia N.D.c). Fourth, firms can shift their business paradigms from short-term interest gains to long-term thinking, ensuring the firm's future (e.g., gaining the trust of consumers and employees) (Taylor 2015).

## 3.2 Green Supply Chain Management in the Textile Industry

Green supply chain management (GSCM) refers to the firms' business positioning integrating environmental thinking into supply chain management (Zhu et al. 2005). GSCM contains product design, material sourcing and selection, manufacturing processes, logistics, and post-usage (e.g., discarding and recycling) (Zhu et al. 2005). Based on GSCM, textile companies can direct their businesses toward an environmentally sustainable approach, so the environment can be protected and the brand image of a firm can be enhanced (Zhu et al. 2005). To execute GSCM within the systematic frameworks, textile companies often adopt global environmental certification systems such as ISO 14001:2015 (environment management system), Blue Sign® Technologies, and OEKO-TEX® Standard 100.

ISO 14001:2015 is one of the most commonly used environmental management standards established by the International Organization for Standardization (ISO), based in Switzerland, consisting of 162 national standards bodies (International Organization for Standardization 2015). ISO 14001 standards provide practical guidance for textile companies to determine and monitor their environmental impact for continuously enhancing their environmental performance (International Organization for Standardization 2015). For instance, in 2011, Adidas Group's headquarters in Germany and its five North American locations (Canton, Carlsbad, Montreal, Portland, and Spartanburg) was awarded the ISO 14001 certificate (Adidas Group 2011). During auditing for the certification, each of Adidas Group locations were assessed according to their GSCM capabilities in identifying, monitoring, and controlling environmentally relevant aspects (e.g., water and energy usage, and waste management) (Adidas Group 2011).

Blue Sign® Technologies is based in Switzerland, and is specifically concentrated on textile production with over 400 partners consisting of textile brands, manufacturers, and chemical suppliers (Blue Sign® Technologies N.D.). It is systemized to approve the textile supply chain's steps involved in chemicals, processes, materials, and products to ensure safety for the environment, workers, and customers (Blue Sign® Technologies N.D.). Since 2000, Patagonia has evaluated their products by using the Blue Sign® Technologies for reducing non-eco-friendly resource consumption of the supply chain as well as managing chemical, dyeing, and finishing processes (Patagonia N.D.a). From raw materials to manufacturing processes, 100 % of Patagonia's base layer products are being produced in accordance with the Blue Sign® Technologies (Fetcher 2012). Based on over 30,000 material assessment reports provided by Blue Sign® Technologies, to execute eco-friendly production processes, Nike controls its global supply chain, including over 800 contracted factories in about 50 countries and hundreds of textile manufacturers (Nike 2013). Together with Patagonia and Nike, Blue Sign® Technologies' fashion brand partners include Adidas Group, Eileen Fisher, G-Star Raw, Lululemon Athletica, REI, and The North Face (Blue Sign® Technologies N.D.).

OEKO-TEX® system is a textile certification system founded in Switzerland that is specialized in textile testing at every production stage including raw materials, intermediate, and end products (e.g., yarns and fabrics in raw and dyed/finished qualities, knits, and ready-made articles) (OEKO-TEX® N.D.a). To be awarded OEKO-TEX® certification, textile companies are required to be tested for harmful substances (e.g., illegal substances, legally regulated substances, non-legally regulated harmful substances, and parameters for health care) for environmental quality assurance (OEKO-TEX® N.D.b). The partners of OEKO-TEX® certification system include approximately 8,500 manufacturers in over 80 countries (OEKO-TEX® N.D.a).

In GSCM approaches, to resolve current environmental issues in the textile industry, it is important that both designers and other stakeholders (e.g., supply chain, merchandiser, and consumer) are aware of their expanded roles involved in the textile product life cycle such as manufacturing, using, and discarding (Khan and Islam 2015).

## 3.3  Eco-design

Aimed at environmental sustainability, eco-design is one of the textile companies' major commitments at product design, the initial phase of the product life cycle (Roy Choudhury 2015). The term 'eco-design' refers to the creation of products in which design and development processes emphasize environmental effects and responsibilities (Lewis et al. 2001; United Nations Environmental Programme N. D.). In the eco-design approach, all possible factors (e.g., ergonomic, environmental, aesthetic, and cost) that might influence the environment are considered to minimize hazardous environmental effects throughout the products' life cycle (Cerdan et al. 2009; Vinodh and Rathod 2010). Thus, during eco-design and product development processes, all decisions and actions must adhere to environmental approaches (Lewis et al. 2001).

Recently, there have been major movements promoting eco-design approaches in the fashion industry by apparel brands and organizations: H&M, Stella McCartney, and the Council of Fashion Designers of America (CFDA). Since 2013, the Swedish fast fashion brand, H&M, has committed to producing some of its products using sustainable textiles made with recycled materials such as plastic bottles, wool, and cotton (H&M 2013). In 2015, H&M also created a $1.14 million grant for the 'Global Change Award,' a global clothing recycling idea competition to promote consumers' environmentally responsible clothing disposal behaviors (Brooke 2015). The luxury brand designer Stella McCartney only uses fake fur or leather for her design, yet the Falabella bag, an iconic design of the brand, is made with a synthetic leather (HBS Working Knowledge 2015). Also, annually, since 2010, in collaboration with Lexus, CFDA has been holding an 'eco-fashion challenge,' a competition to award three entrepreneurial fashion companies contributing to product creation using eco-design approaches (Council of Fashion Designers of America 2014). All these initiatives executed by the apparel companies and organizations contribute to drawing the attention of other apparel companies and consumers to actively participate in environmentally sustainable production and consumption of apparel products.

In the life cycle approach, eco-design can be initiated in different phases of environmental sustainability such as material, manufacturing, packaging, logistics, usage, and disposal (United Nations Environmental Programme, N.D.). Along with these phases, previous studies illustrate three major components of eco-design approaches: (a) material selection, (b) manufacturing process, and (c) rethinking of design.

### 3.3.1 Material Selection

Material selection refers to material phase sustainability in the use of natural materials and production processes that are renewable and biodegradable without incorporating toxic insecticides or fertilizers (Roy Choudhury 2015). To be applicable for eco-design, these natural materials should comprehensively meet the environmental standards in both their properties and processes (e.g., energy usage, material compositions, and disposal) (United Nations Environmental Programme, N.D.). One of the most popular natural materials used in the apparel industry are cellulosic fibers such as cotton, flax, hemp, mulberry, and ramie. Natural cellulosic fibers have composition and biodegradability satisfying the environmental standards and provide superior technical performance (e.g., ventilation, moisture absorption, and natural cooling) and properties (e.g., antimicrobial, moisture wicking abilities, and antiatopic) (Choi et al. 2012; Jang et al. 2015).

### 3.3.2 Manufacturing Processes

In manufacturing processes, instead of using toxic synthetic processes, the application of textile science allows designers to develop new types of eco-friendly materials by integrating natural materials, technologies, and knowledge into smart solutions. In a life cycle perspective, textile science for apparel is an innovative approach to create opulent, distinctive, and personalized experiences through biomimicry and physiological interactions from natural materials and techniques (Van der Veldena et al. 2015). For instance, Jang et al. (2015) incorporated mulberry fiber and titania nanorods (a natural photocatalytic material) to create a new form of textile that has inherent anti-yellowing and antimicrobial properties.

### 3.3.3 Rethinking of Design

Rethinking of design is aimed to meet the needs of both consumers and the environment. Beyond a simple refinement of the current design, refashion considers an interaction between consumers and their clothing in more eco-friendly and innovative ways (McDonough and Braungart 2002). Two common techniques in rethinking of design involve reflecting transformability and adjustable fit in current clothing styles (e.g., detachable outerwear linings, separating zippers, and changing waistlines). Hence, rethinking of design provides benefits for both consumers and the environment with single-pair, multiple-function clothing without requiring the purchase of multiple pairs of clothing.

According to ISO 14062:2002 (the International Organization for Standardization 2002), there are five major benefits of eco-design for consumers and apparel companies. First, eco-design allows cost reduction by optimizing the usage of materials, energy, processes, and waste disposal, as well as reducing costs. Second, it offers the simulations of new products in innovative and creative aspects.

Third, it promotes new product development using material waste or disposed material. Fourth, it augments consumer's purchase intentions beyond their expectations. Fifth, it enhances the organization's or brand's image. Finally, it improves consumer loyalty. Despite these recognized benefits, it is rare to find eco-design approaches in the design of bikes themselves or related gear including apparel and accessories. For instance, the traditional bike helmet is made with expanded polystyrene that is a non-biodegradable petrochemical made of plastic and is difficult to recycle because of the material's minimal weight (Borromeo 2014). Due to the fact that the use of bikes is rapidly emerging as a popular, eco-friendly practice, bike commuters often think that they already contribute to the environment enough and tend not to consider the importance of eco-friendliness associated with their actual bike, cyclist clothing, and accessory usage (Borromeo 2014).

## 4    Five Phases of Environmental Sustainability in the Textile Industry

In 2006, to promote environmentally sustainable product development, the International Organization for Standardization (ISO) released ISO 14040: 2006 (environmental management—life cycle assessment), the improved life cycle assessment (LCA) standards designed to assess the environmental impact of a product throughout its life cycle from raw materials to manufacturing and from consumption to discarding of used products (International Organization for Standardization 2006). These new standards encourage companies to use resources efficiently for reducing negative environmental impacts with the systemized assessment processes over a product's life span (International Organization for Standardization 2006). Based on the ISO standards, in the textile industry, there are five phases in a product's life cycle to be considered for environmental sustainability: (a) material phase, (b) manufacturing phase, (c) retail phases, (d) consumption phases, and (e) disposal phases.

## 4.1    Material Phase

### 4.1.1    The Use of Materials in the Textile Industry

In the textile industry, there are two major categories of materials used to produce textile products, man-made fiber (MMF), and natural fiber. Man-made fiber (MMF) contains synthetic (or artificial) fiber and regenerated cellulosic fibers that are significantly altered in chemical composition, structure, and properties during the production processes (Textile World 2015). In 2009, synthetic fibers account for approximately 65 % of total global fiber production, while natural fibers take up 35 % and over 70 % of synthetic fibers are made from polyester

(Teonline 2009). Through the continuous growth in the textile market, in 2014, total global demands were 55.2 million tons in synthetic fibers, 5.2 million tons in man-made cellulosic fibers, and 25.4 million tons in natural fiber (Textile World 2015).

## Synthetic fiber

Among the synthetic fiber category, nylon, polyester, acrylic, latex, and PVC are major materials that are used in the textile industry. In 1892, the initial invention of MMF was an artificial silk made with cellulosic in France. Based on the textile companies' continuous efforts to develop other cellulosic materials, in the 1930s, nylon, the oldest MMF, was invented and successfully commercialized by DuPont. Nylon is a polyamide made from petroleum taking one of the greatest greenhouse gas inventors (Rydell 2001). During World War II, nylon was mainly used by the military, and later, it replaced silk in the manufacture of women's hosiery. In 2014, 4 million tons of nylon was produced globally and consumed mostly in Asia and the USA (52 % in Asia and 20 % in the USA) (Textile World 2015). In the textile industry, carpet is one of the largest business sectors consuming approximately 17.5 % of total global usage. For apparels, nylon is being used to produce lingerie, sheer hosiery, and swimwear (Textile World 2015). Nylon production emits nitrous oxide, one of the major components of greenhouse gas destroying the ozone layer and 310 times more harmful than carbon dioxide (another greenhouse gas composition) (Greenchoices N.D.). Considering the annual amount of global nylon production, nylon's harmful environmental impacts are significant.

In the early 1940s, polyester was developed by Imperial Chemical Industries (ICI) in England. Polyester is a polymer made from coal, water, air, and petroleum products, and major greenhouse gas emitters (Małgorzata et al. 2003; Rydell 2001). As polyester started partially replacing the demand for nylon, in 1980, global polyester needs were at 5.2 million tons, and by 2000, it had exceeded 19.2 million tons. In 2014, 46.1 million tons of polyester was needed, occupying about 73 % of the total synthetic fiber demand at 55.2 million tons (Textile World 2015).

From production to disposal, one polyester T-shirt emits 5.5 kg of $CO_2$, the same degree of carbon footprint as using 0.6 gallons of gas, burning 5.9 lb of coal, or driving a sedan for 13 miles (see Fig. 1).

Currently, the textile industry accounts for approximately 80 % of polyester production, which creates over 706 million tons of greenhouse gas; this is equal to

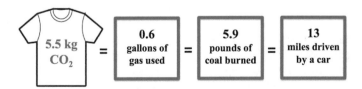

**Fig. 1** Greenhouse gas impacts throughout the life cycle of one polyester T-shirt (Kirchain et al. 2015)

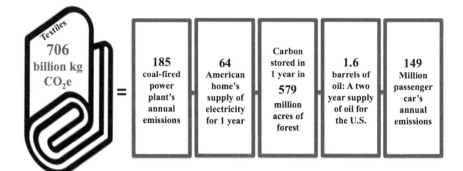

Fig. 2 Greenhouse gas impacts of annual polyester production in the textile industry (Kirchain et al. 2015)

the annual carbon footprint made by burning coal at 185 power plants, the electricity supply for 64 million American homes, carbon storage in a forest of 579 million acres, the US oil supply for two years (1.6 billion barrels), and the annual emissions of 149 million sedans (see Fig. 2). Considering the growth of polyester in the global marketplace, polyester production is anticipated to generate about 1.5 billion tons of $CO_2$ emissions by 2030 (Kirchain et al. 2015).

In 2009, the textile industry accounted for approximately 60 % of total global polyethylene terephthalate (PET) production, and 30 % of that was used to make bottles (Pacific Institute 2009). Annually, 70 million barrels of oil is consumed to produce the virgin polyester for fabric manufacturing (Pacific Institute 2009). Polyester production is a high-energy consumption process requiring a large amount of water and lubricants for cooling, which causes destructive effects to the environment (Greenchoices N.D.). As the fiber is the most demanded across industries, polyester is one of the greatest contributors to natural resource exhaustion and environment contamination.

Acrylic has been widely used to manufacture sweaters, socks, fleece, knitted apparel, and sportswear in the textile industry. During production processes, acrylic, made from a petrochemical material acrylonitrile, generates volatile organic compounds that transform into greenhouse gases (Petrochemicals Europe N.D.). In addition to the greenhouse gas emission, acrylic production consumes 30 % more energy (energy, water, and electricity) than polyester, another high-energy-intensive fiber (Oecotextiles N.D.).

Spandex is an essential material for producing tight-fitting apparel, especially sportswear. Invented in 1958, Spandex is made from over 85 % of the polymer polyurethane consisting of two toxic materials: MDI (methylene diphenyl diisocyanate) and TDI (toluene diisocyanate). These materials make polyurethane very flexible and resistant to the atmosphere for a long time, while causing harmful effects to both the environment and human health (e.g., skin irritation, asthma, lung damage, and respiratory diseases) (American Latex Allergy Association N.D.; Lights N.D.).

Polyvinyl chloride (PVC) is one of the most notoriously non-eco-friendly materials. Since 1960, in the textile industry, PVC has been widely used to manufacture clothing (e.g., protective workwear, outdoor sportswear, and rainwear), shoes (e.g., soles, uppers, and synthetic leather skin), sports equipment (e.g., covering and coating), and bags and luggage (PVC.org N.D.). PVC is made from a petroleum-based toxic plastic and emits greenhouse gases while increasing the human health costs. During manufacturing processes, dioxin and other enduring pollutants are emitted into the environment (e.g., air, water, and land) (Oecotextiles 2014). So far, no safe methods for producing, using, or discarding of the material have been invented to prevent the destructive effects of PVC (Oecotextiles 2014).

### Regenerated cellulosic fiber

Regenerated cellulosic fiber is made by the transformation of natural cellulose to a soluble cellulosic reconstruction in the forms of fiber or a film, thus requiring longer and more complicated manufacturing processes than natural fibers (Alger 1996; Oecotextiles 2012a). In the 1880s, the first regenerated cellulosic fiber rayon was invented as a substitute for silk in France. Rayon had the strong and soft, touchable qualities, similar to silk, while its chemical composition was close to cotton (Board of Intermediate Education Andhra Pradesh N.D.). A traditional viscose process, in which sodium hydroxide and carbon disulfide are treated to transform natural cellulose to viscous liquid, is the most common chemical technique to manufacture rayon (Alger 1996). As the consequence of chemical usage, a traditional viscose rayon process creates negative impacts on the environment when the processed sodium hydroxide and carbon disulfide are drained, untreated, into water sources or emitted into the air, where they are turned into greenhouse gases (Alger 1996: Oecotextiles 2012a). During the weaving process for converting the viscose fibers into fabric, more destructive effects to the environment, as well as human health, will occur with the use of an excessive amounts of chemicals and water (Alger 1996: Oecotextiles 2012a).

### Natural fiber

Natural fibers take up approximately half of the total global fabric consumption (Textile World 2015). The most common natural fibers used in the textile industry include cotton, organic cotton, wool, linen, bamboo, and hemp. In today's textile market, cotton occupies the largest portion (80 %) of global natural fiber usage and accounts for one-third of total global fiber demands (MacDonald 2012). Cotton is a natural material, but still creates problems for the environment and human health. For instance, among entire agricultural crops, cotton is one of the most pesticide- and herbicide-intensive crops, accounting for 16 % of all insecticides and 6.8 % of all herbicides used globally (Greenchoices N.D.; Organiccotton N.D.a). These insecticide and herbicide applications will still remain on the end products after production (e.g., fabrics and garments) and throughout the lifetime of the material and pollute water sources and devastate natural ecosystems by reducing biodiversity in the elimination of the natural enemies of pests (Organiccotton N.D.a).

Compared to other vegetable fiber crops, cotton requires a larger amount of natural resources (e.g., land, water, and electric energy); currently, cotton farms occupy 2.5 % of all global agricultural land. Cotton crops require large amounts of water for irrigation that will, as a result, degrade soil fertility due to salinization. Approximately 60 % of irrigation water was lost in Central and Southern Asia, where major water irrigation channels are available for cotton farms. In the production of industrial fertilizers, cotton accounts for 1.5 % of the total global energy consumption annually, while emitting excessive amounts of the greenhouse gas carbon dioxide (Organiccotton N.D.a).

In addition to the devastation of natural resources, cotton production also contributes to climate change.

For instance, according to the results obtained by the US Environmental Protection Agency's (EPA) greenhouse gas equivalencies (GGE) calculator (the U. S. Environmental Protection Agency N.D.), throughout the product's life cycle, one cotton T-shirt (based on an average weight of 0.4 lb) is estimated to emit about 2.1 kg $CO_2$, or the equivalent of a carbon footprint generated by consuming 0.24 gallons of gas, burning 2.3 lb of coal, and driving an automobile (a sedan) for five miles (Kirchain et al. 2015) (see Fig. 3). Furthermore, massive application of fertilizer nitrates will generate another greenhouse gas nitrous oxide that is 300 times more destructive than carbon dioxide (Organiccotton N.D.a).

The impact of one T-shirt may look small, but the overall cotton's effects in the global textile industry are much more significant. For example, in 2013, cotton was manufactured for 25 million tons globally and around 40 % of these productions were used in the textile industry. Based on the EPAS's GGE calculation results, textile products made with cotton create 107.5 million tons of $CO_2$ emissions, the equivalent of the carbon footprints of annual burning of 25 coal-fired power plants, electricity supply for 13.4 million American homes, annual forest carbon storage of 88 million acres, round-trips to the sun for 1,300 times by a sedan, and two years of carbon emissions from a sedan in New York state (see Fig. 4).

Wool is a fiber obtained from animals like sheep, goats, mohair goats (cashmere and mohair), musk oxen (qiviut), and rabbits (angora) (D'Arcy 1986). Throughout production processes, wool contributes to climate change, from animal breeding to garment mothproofing.

Livestock's belching and gas passing emit methane gas, a source for greenhouse gases that accounts for approximately one quarter of annual agricultural methane

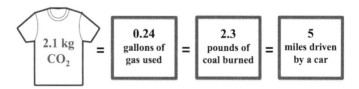

**Fig. 3** Greenhouse gas impacts of one cotton T-shirt (based on an average weight of 0.4 lbs) (Kirchain et al. 2015)

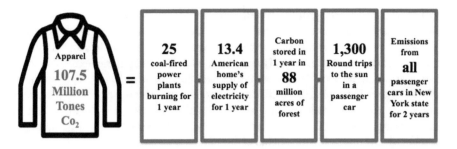

**Fig. 4** Greenhouse gas impacts of annual cotton production in the global textile industry (Kirchain et al. 2015)

emissions; the areas with large livestock populations experience land degradation issues that will cause vegetation change and soil erosion of the land. Water sources are contaminated and exceed suitable drinking and recreational usage levels by fecal matter in livestock farming areas, and the use of toxic chemicals, called the sheep dip, for sheep parasite removal negatively impacts the environment as well as workers (e.g., agricultural and craft workers) (Peta N.D.).

As contrast to cotton and wool produced in traditional methods, hemp and bamboo are natural fibers that do not create environmental problems for two reasons: the first is due to extraordinary productivity, easy cultivation, and high pest tolerance, requiring little or non-agrochemical application, such as pesticides and fertilizers, and second, the deep roots of these materials bind and enrich the soil (Greenchoices N.D.). Bamboo, especially, promotes a range of technical performance capabilities with its natural characteristics and properties (e.g., hypoallergenic, absorbent, fast drying, and antimicrobial) (Greenchoices N.D.). In addition to hemp and bamboo, linen made from flax is also an eco-friendly material that consumes few chemical fertilizers and fewer pesticides than cotton (Greenchoices N.D.).

### 4.1.2 Alternative Materials for Environmental Sustainability

In today's textile industry, to reduce the natural resource consumption and greenhouse gas emission, environmentally sustainable fiber production methods have become available as alternatives to traditional ones in the following textile categories: synthetic fiber (e.g., recycled polyester), regenerated cellulose fibers (e.g., Tencel®, Modal®, and lyocell), and natural fibers (e.g., organic cotton and wool).

*Recycled polyester (rPET)*

Recycled polyester (rPET) has been suggested as an eco-friendly alternative to polyester for two reasons: for one, rPET production requires approximately 33–53 % less energy compared to virgin polyester, and two, rPET generates 54.6 % fewer $CO_2$ emissions than virgin polyester (Libolon N.D.). For instance, since 1993, Patagonia has produced clothing (e.g., Capilene® base layer, fleece, and board short shell jacket) using recycled polyester obtained from recyclable materials such as plastic soda bottles, unusable manufacturing waste, and used garments (Patagonia N.D.b). The use of recycled polyester by Patagonia is an example of a best practice for environmentally conscious manufacturing in the textile industry, providing multiple benefits to the environment and society, such as reducing petroleum usage, increasing landfill life, decreasing greenhouse gas emissions, and promoting polyester-made clothing recycling streams (Patagonia N.D.b). Despite the advantages of rPET, it still creates environmental issues, due to its approximately 30 % greater use of chemical bleaching solutions and water and energy usage for redyeing processes as compared to virgin polyester production (Oecotextiles 2010).

### Modified regenerated cellulose fibers

To reduce the negative impacts of a traditional viscose rayon manufacturing process, the Austrian textile company Lenzing developed Tencel® and Modal®, innovated types of rayon. Tencel® and Modal® adopt more eco-friendly manufacturing processes with the use of non-toxic solvent spinning, which are capable of reducing greenhouse gas emissions and resource consumption as well as recycling water waste (Oecotextiles 2012a). Acetate, triacetate, and lyocell are other types of rayon manufactured by more eco-friendly processes than the traditional viscose rayon process by using natural materials like deconstructed wood pulp or cotton. These rayon family fibers allow soft and silky fabric properties and advanced draping capabilities (Board of Intermediate Education Andhra Pradesh N.D.).

### Organic fibers

Different from traditional fiber cultivation, organic farming methods are suggested based on global organic textile standardizations such as those provided by the Global Organic Textile Standard (multinational), EU (834/2007), the USA (National Organic Program), India (the Indian National Program for Organic Production), and Japan (the Japanese Agricultural Standard) (Global Organic Textile Standard N.D.; Organiccotton N.D.b). These standards provide the systemic approaches concentrated on soil fertility management and crop nutrition by using crop diversification, organic input (e.g., compost, mulch, and manures) and natural pesticide application (Organiccotton N.D.b). In the UK, the Wear Organic project is run by the Pesticide Action Network UK (PAN UK) to reduce the pesticide application problems in cotton as well as create consumer awareness about the environmental impacts of fabric production (Greenchoices N.D.; Organiccotton N.D.b). For organic wool, major environmental advantages come from the eco-friendly farming practices in which it is produced without the use of the toxic sheep dip application (Greenchoices N.D.). In the USA, to get certified as producing organic wool, producers must conform to the federal standards for organic livestock production such as organic feed and forage usage, not use synthetic hormones and genetic engineering application, and not use synthetic pesticides on pastureland treatment (Oecotextiles 2009a).

The use of organic textiles can contribute to reducing global climate change up to 46 %, and natural resource consumption while producing organic cotton may save water by 90 % and energy by 60 % (Ifoam EU Group 2015). Among fashion brands, H&M has continuously been one of the greatest organic cotton users in the global textile industry, together with C&A (Dutch fashion chain) and Puma. In 2013, H&M consumed the largest amount of organic cotton globally, accounting for over 10 % of total cotton used by the company; five percent of this organic cotton usage was from Better Cotton, the nonprofit global organization supporting organic cotton supply chain network (Lamicella 2014). Patagonia is one of the major organic wool market growth drivers in the textile industry. In 2013, the company started to support global organic wool production through a partnership with the Argentine ranchers' network (Patagonia 2015). Since then, Patagonia manufactures almost all of its base layer and sock products by using organic wool (Gunther 2013). See Tables 1 and 2.

## 4.2 Manufacturing Phases

### 4.2.1 Textile Manufacturing Processes

In textile manufacturing processes, such as dyeing, bleaching, and finishing, a large amount of water and toxic chemicals (e.g., benzidine, heavy metals, formaldehyde, and azo) used by textile mills pollutes the environment and becomes the cause of various diseases (allergies, eczema, and cancer) (Oecotextiles N.D.).

*Bleaching*

To be considered suitable material for clothing, the greige material, in its natural form, must be bleached to remove color, odor, and impurities acquired from the use of chemicals for cultivation (e.g., pesticides, fungicides, worm killers, and lubricants). One of the most frequently used bleaching chemicals is chlorine bleach, which is extremely hazardous to the environment and human health. Ecotextiles suggest an eco-friendly bleaching method using hydrogen peroxide made from oxygen and wastewater treatment, so the bleaching processes create less toxic chemical residuals and water waste (Organiccotton N.D.a).

*Dyeing and finishing*

After bleaching, fabrics are dyed for recoloring as well as removing aromatic amines (e.g., benzidine and toluidine). Dyeing processes require the use of multiple toxic chemicals (e.g., heavy metals, pigments, ammonia, and alkali salts). Over 40 % of dyeing colorants include a carcinogen material-bound chlorine, and mordants (color fixer), such as chromium, are extremely toxic and impactful to nature (Oecotextiles N.D.). Additionally, dyeing consumes the largest amount of water in garment manufacturing processes. Currently, the textile mills conduct approximately 3,600 types of textile dyeing procedures using over 8,000 chemicals (Kant 2012). The

**Table 1** Embodied energy used in the production of various fibers (Oecotextiles 2009a)

| Type of fiber | Name of fiber | Energy used in MJ per kg of fiber |
|---|---|---|
| Natural | Organic hemp | 2 |
| | Flax | 10 |
| | Hemp (traditional) | 12 |
| | Organic cotton (India) | 12 |
| | Organic cotton (USA) | 14 |
| | Cotton (traditional) | 55 |
| | Wool | 63 |
| Recycled | rPET | 66 |
| Regenerated | Viscose | 100 |
| Synthetic | Polypropylene | 115 |
| | Polyester | 125 |
| | Acrylic | 175 |
| | Nylon | 250 |

**Table 2** KG of $CO_2$ emissions per ton of spun fiber (Oecotextiles 2009a)

| Name of fiber | Crop cultivation | Fiber production | Total |
|---|---|---|---|
| Polyester (USA) | 0 | 9.52 | 9.52 |
| Cotton (traditional, USA) | 4.2 | 1.7 | 5.89 |
| rPET | | | 5.19 |
| Hemp (traditional) | 1.9 | 2.15 | 4.1 |
| Organic cotton (India) | 2 | 1.8 | 3.75 |
| Organic cotton (USA) | 0.9 | 1.45 | 2.35 |

average-sized mills use about 0.26 million liters of water to dye 8,000 kg of fabric every day; this means each kilogram of fabric consumes approximately 30–50 L of water (Kant 2012). The water wastes created by chemical dyeing are classified as the most hazardous across industries, and they are often drained into water sources without being treated to remove heavy metal-based mordant (Oecotextiles N.D.).

The global denim industry is the largest contributor of dyeing pollution to the environment, as the industry occupies over 30 % of the present-day total global textile industry (Greenpeace 2012). For instance, Xintang, China, the denim capital of the world, manufactures over 260 million pairs of denim apparel annually and accounts for approximately 60 % of total denim production in China and 40 % of total denim sales in the USA every year (Greenpeace 2010). In 2010, Greenpeace tested water samples obtained from throughout Xintang and found five heavy metals (cadmium, chromium, mercury, lead, and copper) in 17 of the 21 samples, with one of these samples containing cadmium that exceeded by 128 times the national limits (Greenpeace 2010).

Natural dyeing methods should be used to minimize the environmental impacts of dyeing processes, though they, however, require equal to double the amount of dyeing materials (e.g., wild plants and lichens) and mordant to the fiber weight while generating lower environmental effects (Oecotextiles N.D.).

## Printing

Traditional textile printing methods refer to the fabric color application processes for creating designated patterns or designs (Kadolph 2007; Oecotextiles 2012b). During the textile printing process, the color is bonded in the fabric properties to be sustained from washing or friction (Kadolph 2007). Traditional textile printing systems vary depending on the use of components facilitating printing process such as wooden blocks, stencils, engraved plates, rollers, and silk screens (Kadolph 2007). Traditional textile printing methods contain many non-eco-friendly aspects, because they consume a large amount of raw materials and energy and create production waste (Oecotextiles 2012b). Considering the significant market size of the textile industry, every year, an excessive amount of toxic chemicals is used and generates hazardous production wastes and environmental effects such as volatile gas emissions to the air. For instance, after each textile printing run, approximately 1.5 gallons of printing paste remains in the printing screen, and equipment that must be cleaned with solvents uses a great amount of water and electricity. These solvents contain toxic chemicals, such as toluene, xylene, and methanol, and water wastes created from solvent cleaning pollute water sources when they are drained, often untreated (Oecotextiles 2012b). In addition, every year, from this volatile toxic solvent usage, every textile printing line of the textile mills emits an average VOC (volatile organic compound) of approximately 14.3 tons for roller and 32 tons for flat and rotary screens (US Environmental Protection Agency N.D.b). In addition to air emissions, these toxic solvent vapors are released throughout the printing manufacturing areas and cause skin and eye irritations to factory workers (Sellappa et al. 2010). Thus, textile designers and company owners need to be aware of the necessity of adopting environmentally sustainable textile printing methods as well as being capable of generating production efficiencies.

## Mercerizing

To enhance luster effects, cellulosic fibers (e.g., cotton, hemp, and linen) can be mercerized after weaving or spinning processes (Oecotextiles 2012c). In addition to luster, mercerization also allows fabrics to experience multiple benefits such as improved moisture absorption (by 7.5–8.5 %), wet resistance, lint reduction, and dyeing methods capable of creating brighter and deeper colors (Beaudet N.D.; Theballofyarn N.D.). In particular, dyeing capabilities can be enhanced by 25 % and dyed colors stay longer after a fabric is mercerized (Beaudet N.D.). Despite these benefits, the textile mercerization creates negative environmental impacts, the results of the use of toxic chemical sodium hydroxide (salts) (Beaudet N.D.). Most cotton fabric production wastewater contains 2,000–3,000 ppm of salt concentrations that significantly exceed the federal limit of 230 ppm; hence, the US Environmental Protection Agency (EPA) restricts textile mills from draining mercerization water waste that includes sodium hydroxide into water sources (Aquafit4news 2011; Radha et al. 2009).

*Assembly*

Assembly process refers to gathering each garment component into a finished garment in a workstation setting, and is a designed task performed according to the protocols (Chan et al. 1998). One of the most labor- and material-intensive parts of assembly is the sewing process that involves connecting components together; thus, in order to save unnecessary material wastes and decrease mistakes, entire work assembly processes should be simulated to prioritize work orders before executing the actual sewing process. Advanced planning of assembly helps to enhance the overall production quality (Cooklin 1991). The garment assembly process requires fluent knowledge and skills of garment production, as well as many collaborative processes and schedule arrangements (Cooklin 1991). Due to this complicated structure, garment assembly is difficult to automate with computer technology (Chan et al. 1998). In addition, during the assembly process, a significant amount of electricity is consumed for running equipment, such as sewing machines and pressers, in which it will later become the cause of greenhouse gas emissions (Palamutcu 2010). Therefore, to reduce the potential environmental impacts from the assembly process, designers and manufacturers must consider using alternative assembly approaches that allow saving labor, material, and energy usage.

*Packaging*

Packaging is a process that integrates wrapping around a product for multiple purposes such as containing, protection, identification, and promotion, assisting in making the product more marketable (Entrepreneur N.D.). There are several requirements for product packaging: first, it should be durable with consideration to shipping and stocking in the warehouse as well as displaying in retail stores; second, as the face of the product, packaging must illustrate specific information about the product (e.g., instructions about how to handle or use); and third, it should be designed in such a way that it can be easily opened and re-closed exclusively by the purchaser (Entrepreneur N.D.). For filling these requirements, synthetic packaging materials are often used for the packaging of textile products that are not eco-friendly (Szaky 2014). A polybag is a common form of textile product packaging made from polyethylene (PET) that is hard to be recycled or is not biodegradable. Due to the textile industry's standards for individual product wrapping, in the USA, 3.8 million pounds of PET wastes is created and only 12 % of these wastes are recycled every year (Szaky 2014). Hence, packaging must be innovated, so it is not landfilled after a single use, or should be biodegradable or recyclable, so negative environmental impacts of packaging can be reduced.

### 4.2.2 Eco-friendly Alternatives to Traditional Manufacturing Processes

Based on the use of innovative technologies, there are eco-friendly alternatives that are being used by textile companies to resolve environmental issues created by traditional manufacturing processes. Among these innovative technologies, environmentally sustainable alternatives include those for bleaching (e.g., air dyeing, less water wet finishing, laser bleaching and washing, and ozone bleaching and finishing), printing (e.g., digital printing technology), mercerization (e.g., electrochemical cell mercerization), and assembly (e.g., zero-waste garment design, seamless knitting technology, and 3D-integrated design technology).

## Air dyeing

Air dyeing technology (ADT) is a dyeing method using air rather than water (Kant 2012). ADT allows textile companies to still produce aesthetically appealing colors of garments without causing destruction to the environment. ADT creates approximately 84 % fewer greenhouse gas emission levels and consumes 87 % less energy, and with a single fabric piece, at one time, its different fabric sides can be dyed in different colors, so unique designs can be created (Kant 2012). ADT has been used to dye various fashion products, such as swimsuits and rPET (recycled PET), and by fashion designers in major design houses in New York (Kant 2012).

## Less water wet finishing

Levi's is one of the world's leaders in promoting wet-finishing methods that consume less water. Since 2011, the denim giant has produced the Water<Less jeans, an innovative technique to create denim finishing effects using much less water than that required by traditional wet dyeing and finishing processes (Elks 2014; Levistrauss N.D.). To create its unique denim appearance, these traditional processes require three to ten spins of washing on average. Water<Less technology, however, combines the multiple wash cycles into a single process (Elks 2014). In addition to this, the company has integrated ozone processing to create the washed-down effects to the garments (Elks 2014). As a result, each finishing process in Levi's Water<Less jean's method reduces water consumption by approximately 96 %, and since 2011, the cumulated water saving is over 770 million liters, which corresponds to the amount of drinking water for 811,000 people (Levistrauss N.D.).

## Laser bleaching and washing

A laser (an acronym for light amplification by stimulated emission of radiation) used in the industries refers to an instrument that generates light through an optical amplification process in accordance with the electromagnetic radiation's stimulated emission (Hecht 2005). A laser instrument generates an extreme amount of heat to melt a small focused area for changing a material's appearance (Hecht 2005). In the textile industry, the $CO_2$ laser instrument is commonly used, due to its capabilities in creating designed surface patterns or effects more precisely and within a short time with minimal damage to the textile properties (Bertolotti 2015).

As an alternative to synthetic bleaching and washing methods, apparel designers can use laser technology to execute elaborate regional effects, such as the worn-out look, whiskers, and design motifs/patterns, without using water and chemicals. The use of lasers engineered by computer systems allows designers to create more accurate and repeatable bleached and washed effects (Apparel 2014). One of the shortcomings of the laser method is the high cost for an individual piece of equipment's operation, and side by side at each piece, it provides limited efficiencies for overall bleaching (Apparel 2014).

## Ozone bleaching and finishing

The use of ozone technology allows textile companies to create natural beaching effects by using ozone gas instead of chemicals (Apparel 2014; Khalil 2015). In ozone finishing methods, oxygen ($O_2$) is changed to ozone gas ($O_3$) through a series of processes such as dampening, sun exposure, and rinsing of denim garments; after that, $O_3$ is changed back to ordinary, harmless oxygen before emission into the environment (Apparel 2014). Ozone technology is capable of reducing the consumption of water, energy, and dyestuffs (e.g., chemical, enzyme, and stone). For instance, Levi's is currently a leader of ozone technology adoption in the textile industry (Elks 2014). Ozone requires two to three times the washing processes only, while traditional chemical washing necessitates six to seven times (Khalil 2015). Additionally, ozone takes 15 min for bleaching, while traditional chemical methods require 30–45 minutes (Apparel 2014; Khalil 2015). Thus, it contributes substantially to reducing hazardous environmental impacts compared to traditional wet, chemical bleaching, and finishing processes.

## Digital printing

Due to the fact that textile printing embraces non-eco-friendly aspects requiring a large amount of raw materials, energy, and generating production waste, digital printing technology emerged as an alternative to conduct printing while reducing harmful wastes to the environment (Oecotextiles 2012a). Digital printing allows multiple benefits to the textile companies and environmental protection such as reduced material and energy usage, enhanced process efficiency, and a longer equipment life span (Boyd 2014; SPGprint N.D.). More specifically, a digital printer consumes an approximate additional 5 % of textile materials and ink, while a traditional screen printer requires about 15 % of these consumable materials (Boyd 2014; SPGprint N.D.). The digital image distribution function makes use of physical materials, such as a traditional silk screen, to make print-on-demand possible; hence, additional costs can be avoided for handling physical materials (e.g., transportation, warehousing, and waste disposal) (Boyd 2014). Based on the computerized simulation, a high-quality image can be customized with precision requiring a reduced correction process and energy consumption (SPGprint N.D.). Traditional screen printing methods generate a large amount of toxic chemical waste, since a printing plate is made from polyester or rubber, as well as toxic solvents for roller ink removal (Boyd 2014). Even though digital printing is still involved in some chemical usage, it requires a reduced amount of less harmful solvents for ink removal (Boyd 2014; SPGprint N.D.). In sum, digital printing contributes more to preventing negative impacts on the environment by reducing physical and chemical material waste creation than traditional printing methods.

## Electrochemical cell mercerization

To replace traditional sodium hydroxide-based mercerization, electrochemical cell mercerization can be used to create mercerized effects without incorporating neutralizing acids or bleaching materials into the processes. Electrochemical cell mercerization is a process for creating an oxidation that is an electric current flow reaction between a solid electrode and substance by using electrical energy (Theballofyarn N.D.). Electrochemical methods, however, demand more production costs than traditional ones, due to special equipment requirements (Radha et al. 2009; Theballofyarn N.D.).

## Zero-waste garment design

Zero-waste design is an eco-friendly design approach used to minimize a fabric scrap from the cutting room. To reduce waste in pattern cutting, a designer fits the puzzle of each clothing pattern component (e.g., gussets, pockets, collars, and trims) together or generates a drape pattern directly onto a mannequin (Rosenboom 2010). The zero-waste pattern cutting approach could reduce millions of tons of fabric waste every year, because the textile industry normally generates 15–20 % of fabric materials leftover from the clothing manufacturing process that are going to be landfilled (Rosenboom 2010). Zero-waste design techniques have been developed by fashion designers, especially those in New York, by using innovative design techniques (e.g., digitized pattern design, WholeGarment® knitting technology), and the textile industry adopts some of these designs for commercialization (Rosenboom 2010).

For instance, in the spring of 2011, Parsons the New School for Design initiated an environmentally sustainable design project in collaboration with design students, industry partners, and the Parson Milano campus. This project concentrates on the zero-waste pattern cutting, as well as the garment life cycle extension and innovative use of natural resources. An eco-friendly fashion brand Loomstate commercialized a selected student collection from this project in the fall of 2011 (Parsons School of Design 2011).

Similar to these Parsons' initiatives, in the textile industry, a luxury fashion brand Issey Miyake has made another distinctive example of zero-waste design. In 1976, designer Miyake initiated his a piece-of-cloth (A-POC) design concept to make a garment in a single piece covering the entire body, and since 1999, the designer has presented creative collections using this concept (Isseymiyake N.D.). In 2010, the Issey Miyake brand started another innovative zero-waste design approach inspired by origami, a Japanese decorative paper folding art (Hethorn and Ulasewicz 2015; Isseymiyake N.D.). The origami design concept allows designers to create 3D geometric forms from a flat surface by using a computer program initially developed by a Japanese computer scientist Jun Mitani (Hethorn and Ulasewicz 2015; Isseymiyake N.D.). Based on the zero-waste design concepts and techniques, Issey Miyake is an outstanding example of the textile companies that have experienced huge success not only in terms of economic profits, but also in terms of environmental protection. Aside from the case of Issey Miyake, the zero-waste design techniques, however, still hold some limitations for mass production by large textile manufacturers, due to the increased costs and reduced efficiencies to execute elaborated processes required because of the existing infrastructure (Hethorn and Ulasewicz 2015). Besides the cost and efficiency issues, an environmentally sustainable design approach, such as zero-waste design, must be continuously experimented with to develop feasible techniques to be adopted by the textile industry.

# 5   3D Seamless Knitting Technology

3D seamless knitting is an alternative to traditional knitting techniques that require knitting of each part of garment separately and then cutting and sewing these parts together afterward, a process in which approximately 30 % of materials are lost (David 2015; Ecofashiontalk 2012; Textilelearner N.D.). Quite different from traditional knitting techniques, 3D seamless knitting technology allows minimizing waste through knitting the entire garment in one process without seams in a three-dimensional construction on the knitting machine, thus reducing labor and material loss associated with cutting and sewing processes (David 2015; Ecofashiontalk 2012; Shima Seiki N.D.a). Additionally, due to the seamless nature, garments knitted by WholeGarment® provide advantages, such as being soft, lightweight, and stretchy, and add mobility that enhances the wearing comfort (Shima Seiki N.D.a). It should be noted that in this context, the term 3D seamless knitting is different from traditional seamless garment production using circular knitting machines in which stitching is required to produce the end product. There are two major 3D seamless garment knitting machine manufacturers in the global apparel market: Shima Seiki and Stoll (David 2015; Shima Seiki N.D.a).

Based in Japan, Shima Seiki Mfg. Ltd. has produced a 3D seamless knitting machine called WholeGarment® knitting technology for the past fifty years (Shima Seiki N.D.a). Shima Seiki's recent developments provide apparel designers more elaborated and complex 3D seamless knitting technology (Shima Seiki N.D.a). Some major customers of Shima Seiki's WholeGarment® knitting machine are Max Mara (a fashion designer brand) and Martignoni Paola (an Italian knitwear manufacturer) in the global textile industry (Hunter 2015; Shima Seiki N.D.a). The SDS-ONE APEX3 software is connected to the WholeGarment® knitting machine that provides functions for designers to simulate style and assembly precisely in 3D before starting the actual production process, preventing unnecessary material waste or mistakes (Shima Seiki N.D.a). In 2004, Shima Seiki was awarded ISO 14001:2004 (environment management system) certification as one of the key leaders of environmental sustainability in the global textile industry (Shima Seiki N. D.b). The company established a solar-powered generation system on a large scale to supply electricity to the factories that produce the WholeGarment® knitting machines, which account for 10 % of the company's annual electricity consumption (Shima Seiki N.D.b).

Founded in 2004, SAEM Collection Ltd. is one of the largest knitting garment manufacturers of South Korea whose business partners include major fashion brands in Korea and China. As one of the biggest accounts for a European 3D seamless knitting machine producer in Korea, SAEM Collection Ltd. has 12 3D seamless knitting machines at its production facility producing over 60,000 pieces of 3D seamless knitting garments annually (SAEM Collection Ltd. 2016).

Since 2009, SAEM Collection Ltd. has invested in 3D seamless knitting for two reasons: first, the use of the 3D virtual design simulation of 3D seamless knitting technology reduces 10–15 % of material wastes during the production process and allows the company to save costs of approximately US $ 0.12 million annually (worth 30 tons of waste materials that would be landfilled); and second, seamless knitting's high product quality enhances consumers' satisfaction and the image of the company (SAEM Collection Ltd. 2016). See Figs. 5–8 for the SAEM's 3D seamless knitting design process.

3D-Integrated Design Technology

In today's textile industry, 3D-integrated design technology is a computer platform that allows designers to accomplish 3D scanning, modeling, and visualization for elaborate and accurate virtual simulation on the body (Apparel 2016). 3D-integrated design technology contributes to screening garment design specifications and reducing potential construction errors and the number of physical samples made; thus, fewer resources will be consumed throughout manufacturing processes (McGregor 2015). Furthermore, designers and merchandisers can see 360 degree prospective product features several months before a physical sample is made (Apparel 2016; McGregor 2015) (see Fig. 9).

In addition to the design process, 3D-integrated design technology can also be used for enhancing consumers' in-store retail shopping experience with virtual garment fitting visualization capabilities that are realistic and interactive for consumers (McGregor 2015). Some large 3D-integrated design technology providers are Gerber Technology, Optitex, and Lectra.

Optitex, based in New York, is one of the industry's leaders in providing an integrated 2D CAD and 3D digital product solution for textile manufacturing and

**Fig. 5** SAEM Collection Ltd.'s 3D seamless knitting design created by the 3D seamless knitting software

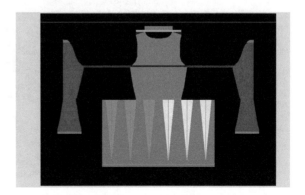

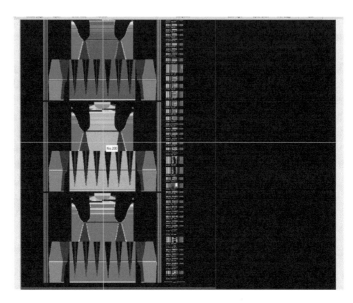

**Fig. 6** SAEM Collection Ltd.'s 3D seamless knitting design created by the 3D seamless knitting software

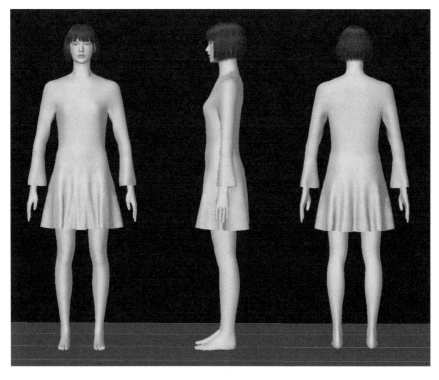

**Fig. 7** SAEM Collection Ltd.'s 3D seamless knitting design created by the 3D seamless knitting software

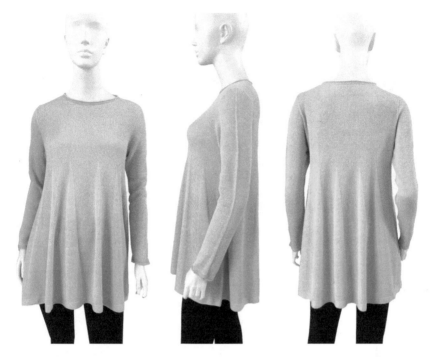

**Fig. 8** SAEM Collection Ltd.'s 3D seamless knitting design created by the 3D seamless knitting software and machines of the factory in South Korea (SAEM Collection Ltd. 2016)

retail companies. Some of the Optitex customers include Roberto Cavalli, Scott Sports, and Under Armour (Optitex 2016). Optitex continuously puts efforts into creating innovative products and solutions to provide its consumers with a more personalized experience. This includes maintaining a standardized, consistent quality in its technology for product development, manufacturing, and marketing (Optitex 2016). See Figs. 9 and 10 for the 3D visualization image examples created by Optitex software.

## *Packaging*

As packaging of textile products makes significant contributions to the increase of landfills, there is an increasing number of apparel companies that use eco-friendly packaging in the global textile industry. For instance, in 2010, Puma presented the Clever Little Bag in collaboration with a design company FuseProject (Puma 2010). The Clever Little Bag combines the shoebox and bag without using additional tissue paper or shopping bags and can be used as a shoe container as well as an eco-bag (Puma 2010). The Bloomingdale's department store uses shopping bags called Brown Bag made with 100 % recyclable material and natural craft methods (Macysgreenliving 2013). The department store giant also sells drinking water boxes made with recyclable paper; 10 % of the sales profits are donated to the

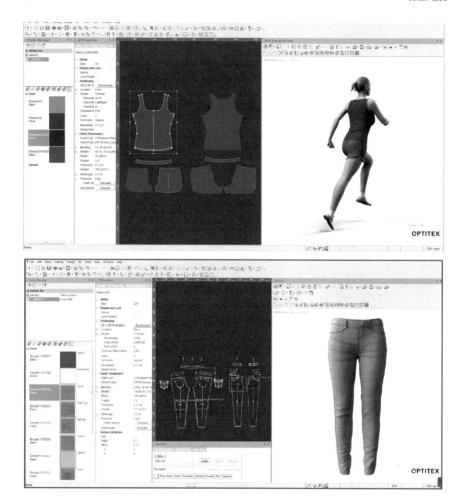

**Fig. 9** 3D visualization images and digitized pattern works created by Optitex software (Optitex 2016)

**Fig. 10** 3D visualization of a jacket constructed with Optitex and actual garment sample image (Lee 2016)

World Water Relief Foundations, and an additional 10 % of the profits are used for reforestation foundations (Macysgreenliving 2013). In 2011, the denim brand Lee presented the Never Wasted shopping bag designed to be transformed for the reuse in different ways (e.g., as a calendar, message tag, envelope, and container).

## 5.1 Retail Phases

### 5.1.1 New Business Model for Environmentally Sustainability

It is important that environmental sustainability is considered in the retailing phase in the global textile industry. For the retailing phase, it is essential that textile companies have a new innovative business model committed to the environment that embraces the company and stakeholders and measurement standards for how to measure and track the company's practice in environmental sustainability (Clinton N.D.). There is growing attention to an innovative business model geared toward environmental, social, political, and technological aspects as future growing and success strategies of the firms (Clinton N.D.). Greentailing, also called eco-tailing, is one of the most common business models dedicated to environmental sustainability.

#### Key market trends in the retail industry

In today's retail market, there are several key market trends directing retailers' success and long-term growth: (a) greentailing, (b) demographic shifts, (c) experiential marketing growth, (d) thinking out of the box, and (e) selling service, not just products.

First, greentailing refers to a business practice committed to environmentally conscious, socially responsible, and economically profitable retailing for all business activities (Perera 2009). It considers the effects to retailers' dedication to the environment and society, consumer awareness and behavior, and employees and the supply chain, as well as returns to shareholders (Perera 2009). Greentailing caters to the growing demands of consumers seeking organic, environmentally and socially sustainable, and well-being products. To respond to these demands, greentailing requires the supply chain and retailers to become capable of implementing ideas and increasing profits while meeting eco-friendly standards in every aspects of the business (Perera 2009; Stern and Ander 2008). Greentailers whose major business emphasis is on greentailing grow faster then competitors with superior product innovation capabilities, because of their environmental sustainability. For instance, due to endless green evolutions, two leading greentailers, Walmart and Whole Foods, constantly grow much more quickly compared to their competitors who are non-greentailers (Stern and Ander 2008). Second, today's major shifts in consumer demographics alter market characteristics that are different from five years ago (Zadro 2014). Due to aging baby boomers' fading out of retail consumer markets, consumer demographics have become younger than previous market. The resulting emergence of technologies is now an essential part of consumer concerns (Zadro 2014). Third, consumers' in-store experience and their engagement in the brand are new powerful drives to creating sales and profits beyond the price of the products (Stern and Ander 2008). Fourth, retailers must think out of the box to discover novel and innovative ways to connect to consumers, especially in consideration of non-traditional consumers (e.g., online shoppers) (Stern and Ander 2008). Finally, today's consumers do not want to buy mere products, but expect service throughout their shopping experiences (Stern and Ander 2008).

The success of today's retail business heavily depends on how to control consumers.

Therefore, in today's retail industry, retailers must understand any variables that may affect consumers' decision-making as related to these key market trend changes. Retailers should create unforgettable impressions for their target consumers through ceaseless attempts linking to their demands based on the innovative strategies in merchandising, marketing, and packaging of products (Zadro 2014). To that end, greentailing approaches are impactful present-day drivers of retailers' growth.

### *Greentailing*

In greentailing, throughout stores, products, and people, corporate social responsibility (CSR) plays a significant role in communication with consumers (Stern and Ander 2008). Hence, greentailers must sell products composed of green elements that are organic and natural and do not include toxic ingredients, are locally grown and ethically sourced through fair trade, and, most importantly, are environmentally sustainable (Stern and Ander 2008). In the meantime, greentailers must provide profit returns to their stakeholders (e.g., supply chain, stores, and employees) through socially and environmentally responsible business practices (Stern and Ander 2008).

There are two examples of eco-retailers in the textile retail industry (Academy Sports + Outdoors and Samsung C&T) that have committed to environmentally and socially responsible fashion business practices. Both of these companies provide evidence that greentailing contributes to building a company's positive image as well as increasing economic profit gains to the stakeholders. The Texas-based sports brand specialty retailer Academy Sports + Outdoors is one of the leaders of greentailers due to the company's commitment to conform with California's Proposition 65 and California Transparency in Supply Chains Act (SB 657). California's Proposition 65 has established a list of over 850 toxic chemicals impactful to humans and nature and restricts the use of these toxic chemicals in the manufacture of products (Academy Sports + Outdoors N.D.). Throughout the company's private label manufacturing processes, Academy Sports + Outdoors constantly monitor all of its supply chain to ensure adherence to California's Proposition 65 and California Transparency in Supply Chains Act (SB 657) (Academy Sports + Outdoors N.D.a, N.D.b).

Established in 1954, Samsung C&T is a large, global, apparel design and retail company with annual sales that exceed two billion dollars (USD). The company operates as an affiliate of Samsung Group and currently manages over 50

**Fig. 11** Samsung C&T's the HEARTIST store in Seoul, South Korea (Samsung C&T 2016)

companies' fashion brands, including imported, licensed, and private labels (Samsung C&T 2016). Based on the company's strong commitment to corporate social responsibility (CSR), in 2013, to celebrate Samsung C&T's 60th anniversary, the company's CSR store called the HEARTIST HOUSE was established. The name HEARTIST is a compound word of "heart" and "artist" that represents the symbolic meanings of the store to reflect eco-friendliness, sharing, and donation. The HEARTIST is designed for a shopping experience similar to donation and with five principles: reduce, reuse, recycle, refine, and recover. There are five distinctive commitments of the HEARTIST to environmental sustainability. First, the products sold in the store can be grouped into three categories: (a) eco-friendly items made with organic and natural materials, (b) donated clothing items from the company's major luxury fashion brands (e.g., Bean Pole, Galaxy, Rogatis, Kuho, LeBeige), and (c) upcycled clothing items created by young contemporary fashion designers who have donated their talents. Second, the store's shopping bags are made from biodegradable, recycled plastic materials and designed for a shareable bag that allows a consumer to bring it back to the store after he or she fills it up with items for donation. Third, a warehouse built in the 1940s was converted to the flagship store through the use of minimized synthetic materials. Fourth, over 50 % of the fixtures used in the store are made with recycled materials. Fifth, the cooling water used for the store's air-conditioning systems is recycled to nourish the flower garden. Additionally, the company's employees, including fashion designers and merchandisers, donate their labor to work for the store as the store staff members. Samsung C&T donates a half of the profits gained from the HEARTIST store to charities and organizations that promote environmental sustainability through donations, campaigns, and events. See Fig. 11 for Samsung C&T's the HEARTIST store images.

## 5.2 Use Phases

Consumption of fashion products is affected by consumers' desire for self-expression and identity creation (Berger and Heath 2007). Due to the significant consideration of the individual identity created by fashion styles, as well as a lack of knowledge of clothing's harmful environmental effects, consumers often care less about ethical or sustainable clothing consumption (Berger and Heath 2007). There are two different trends that affect consumers' apparel consumption behaviors in today's apparel market: fast fashion and slow fashion.

### 5.2.1 Fast Fashion

Fast fashion refers to apparel companies' business strategy to manage the supply chain efficiently to provide the newest fashionable products quickly responding to consumers' demands (Levy and Weitz 2008). As a result, fast fashion's time-to-market (length of time in the product development process required from design ideation to the finished product) is only a few weeks, much shorter than the standard six-month time-to-market in the apparel industry (Tokatli 2008). For instance, the European fast fashion retailer Topshop reduced the brand's time-to-market to six to nine weeks and H&M's time-to-market is only three weeks (Siegle 2011). In addition to a reduced time-to-market, producing a large number of styles of clothing is another distinction of fast fashion retailing. For example, the largest global fast fashion retailer, Zara, produces 12,000 styles selected from 40,000 styles created by 200 in-house designers annually (Siegle 2011). Because of the characteristics of fast fashion in providing the newest fashion styles at low prices in quick turnover cycles, fast fashion has created a disposable clothing culture in which consumers throw away clothing after several times of use; such a consumption culture increases waste material amounts to be landfilled due to a reduced product life cycle (Birtwistle and Moore 2007).

### 5.2.2 Slow Fashion

As a consequence of the increasing environmental impacts created by clothing consumption (especially fast fashion), consumers' environmental awareness is growing, increasing a niche for slow fashion products that promote ethical clothing consumption (Deirdre and Riach 2011). Apparel products manufactured in eco-friendly ways allow consumers to make ethical and consumption choices and create an environmentally conscious identity (Joy et al. 2012).

In 2008, the term *slow fashion* was first introduced by a sustainable design consultant Kate Fletcher as an opposing approach to fast fashion (Phelan 2012). Slow fashion refers to timeless apparel that lasts a long time and is not affected by rapidly changing fashion trends (Fletcher 2010). Beyond mere adoption of organic materials, slow fashion includes environmentally sustainable fashion usage based on consumers' environmental awareness of impacts generated throughout the entire life cycle of textile products (Phelan 2012).

In environmentally sustainable business approaches, slow fashion companies focus on product durability and reusability when designing their clothing (Fletcher 2010). Hence, slow fashion trends allow consumers to purchase timeless designs that can last a long time and maintain a high product quality and encourage designers to create seasonless sellable products over time (Adamczyk 2014; Phelan 2012). These new slow fashion products appeal to consumers seeking unique styles with a willingness to pay premiums; hence, slow fashion can reduce the textile industry's carbon footprint through subtle alternative ways without overloading environmental pressures just on textile companies (Adamczyk 2014). As a leading example of a slow fashion brand in the textile industry, Eileen Fisher presents timeless and ageless designs by reproducing designated icon styles over seasons while catering to new trends for a broad consumer range (Hill 2014). Since 2009, Eileen Fisher has executed a recycling program collecting about 300,000 clothes to be sold at the company's eco-friendly concept stores called Green Eileen stores (Kinosian 2015).

### 5.2.3 Future of Clothing Consumption

As the market size of the fashion industry grows, a large amount of clothing disposal causes serious environmental problems by requiring more and more landfill space every year around the world (Siegle 2011). Therefore, to promote environmental sustainability, a more holistic approach is needed to change consumers' clothing consumption behavior to encompass concerns or responsibility to the environment and society. Together with consumers, apparel companies also have to rethink their green business practices in terms of how clothing products are designed, manufactured, consumed, and disposed of for protecting the natural environment.

## 5.3 Disposal Phases

After consumers use textile products (e.g., apparel, footwear, and accessories), these used textile wastes are often discard. Every year, there is an increasing amount of textile waste created globally, especially in the world fashion capitals like the USA and UK. In the USA, textile production (e.g., apparel, footwear, and accessories) has continuously increased. In 1999, approximately 18.2 billion pounds of textiles were produced. By 2009, this amount increased by 25.46 billion pounds, and by 2019, it is expected to grow by 35.4 billion pounds (Council for Textile Recycling N.D.). Based on 2009's textile production amounts of 25.46 billion pounds, every US resident consumed around 82 lbs of textile products per year, meaning that currently, American people purchase clothing five times more than in the 1980 (Cline 2014; Council for Textile Recycling N.D.). After these textiles were used, only 3.8 billion pounds of the textile wastes, that account for 15 % of total global landfill waste, was donated or recycled; the remaining 21 billion pounds of textiles (85 %) were landfilled (Cline 2014; Council for Textile Recycling N.D.). Similar to the USA, in the UK, it is estimated that 10 % of the 1.4 million tons of used textile wastes is landfilled every year (King 2012). Due to a large amount of textile wastes created annually, the UK's recycling and waste management sectors are faced with challenges of continuous landfilling in the future (King 2012).

As a result of the growing interest in green clothing consumption movements, discarded used textile products are currently recognized as an economic value regenerator through reuse and recycling (Cline 2014). In fact, over 90 % of those textile wastes are recyclable. For example, if textile wastes are recycled, among these waste materials, 45 % can be worn as secondhand clothing; 30 % are able to be cut up and produced as industrial rags; 20 % is biodegraded after the landfill; and only 5 % is unusable (Cline 2014). Thus, in a broad point of view, these discarded materials have high potential to generate revenue for local authorities and waste management companies (King 2012). Recycling provides multiple benefits to consumers: reduction of the need for new apparel manufacturing, reduction in manufacturing, which saves huge energy sources as well as raw materials, and decrease in energy and raw materials, which creates less environmental pollutants (SecondHand 4 Business Ltd. N.D.). Therefore, consumers should be encouraged to participate in environmentally sustainable clothing disposal practices such as recycling and donation.

# Appendix

| Category | Name | Contact | Phone | Location | Web site |
|---|---|---|---|---|---|
| Branded company | SAEM Collection Ltd. | CEO/Kim, In-Kyung | 82-2-478-4575 | Seoul, South Korea | N/A |
| | Samsung C&T | P.R. Team Director/Shim, Moon-Bo | 82-70-7130-9114 | Seoul, South Korea | www. samsingfashion.com |
| Software company | Optitex USA | MarCom Manager/Emma Datny | 212-629-9053 | New York, USA | Emma. Datny@Optitex.com |
| 3D seamless knitting equipment company | Shima Seiki USA | Sales Team | 212-629-9053 | New York, USA | www.shimaseikiusa. com |

# References

Abreu MCS (2015) Perspectives, drivers, and a roadmap for corporate social responsibility in the textile and clothing industry. Text Sci Clothing Technol 1–21

Academy Sports + Outdoors (N.D.) California Proposition 65. http://www.academy.com/shop/en/store/california-proposition-65. Accessed 13 Feb 2016

Adamczyk A (2014) Why brands and retailers are running with the 'Slow Fashion' movement. http://www.forbes.com/sites/aliciaadamczyk/2014/11/20/why-brands-and-retailers-are-running-with-the-slow-fashion-movement/#2dfc8ae65059. Accessed 13 Feb 2016

Adams CA, Frost GR (2008) Integrating sustainability reporting into management practices. Acc Forum 32(4):288–302

Adidas Group (2011) ADIDAS group environmental management awarded with ISO 14001 certificate. http://www.adidas-group.com/en/media/news-archive/sustainability-news/2011/adidas-group-environmental-management-awarded-iso-14001-certificate/. Accessed 2 Mar 2016

Alger M (1996) Polymer science dictionary, 2nd edn. Chapman & Hall, London

American Latex Allergy Association (N.D.) Cotton, Nylon, Spandex and Allergies. http://latexallergyresources.org/articles/cotton-nylon-spandex-and-allergies. Accessed 2 Mar 2016

Apparel (2014) Ozone finishing for denim reduces environmental impact, processing costs and processing time. http://apparel.edgl.com/news/Ozone-Finishing-for-Denim-Reduces-Environmental-Impact,-Processing-Costs-and-Processing-Time94272. Accessed 13 Feb 2016

Apparel (2016) New IEEE Industry Group to focus on immersive shopping experiences. http://apparel.edgl.com/news/New-IEEE-Industry-Group-to-Focus-on-Immersive-Shopping-Experiences103981. Accessed 9 Feb 2016

Aquafit4news (2011) The Textile Industry in AquaFit4Use. http://www.aquafit4use.eu/userdata/file/Newsletters%20and%20leaflets/11%20Aquafit4news%20September%202011.pdf. Accessed 2 Mar 2016

Beaudet T (N.D.) What is Mercerized Cotton? http://fiberarts.org/design/articles/mercerized.html. Accessed 5 Mar 2016

Berger J, Heath C (2007) Where consumers diverge from others: identity signalling and product domains. Journal of Consumer Research 34:121–134

Bertolotti M (2015) Masers and lasers: an historical approach, 3rd edn. CRC Press, Boca Raton, Florida, pp 89–91

Blue Sign® Technologies (N.D.) Bluesign technologies ag – Global partner of a sustainable textile industry. http://www.bluesign.com/home/about-us#.Vu-CYYd_dds. Accessed 2 Mar 2016

Board of Intermediate Education Andhra Pradesh (N.D.) Classification and general properties of textile fibres. http://bieap.gov.in/Pdf/CGTPaperII.pdf. Accessed 2 Mar 2016

Borromeo L (2014) Cycling is a green activity but finding sustainable, ethical cycling gear is hard. Retrieved from http://www.theguardian.com/sustainable-business/sustainable-fashion-blog/cyclists-sustainable-ethical-cycling-gear-helmets-saddle. Accessed 1 Mar 2016

Boyd A (2014) Go digital & go green: the environmental benefits of digital labels. http://www.bluelabeldigital.com/go-digital-go-green-the-environmental-benefits-of-digital-labels/. Accessed 15 Feb 2016

Birtwistle G, Moore C-M (2007) Fashion clothing—where does it all end up?". Int J Retail Distrib Manag 35(3):210–216

Brooke E (2015). H&M creates a €1 million grant for innovations in clothing recycling. Retrieved from http://fashionista.com/2015/08/hm-recycling-grant. Accessed 4 Mar 2016

Cerdan C, Gazulla C, Raugei M, Martinez E, Fullana-i-Palmer P (2009) Proposal for new quantitative eco-design indicators: a first case study. J Clean Prod 17(18):1638–1643. Accessed 4 Mar 2016

Chan K-C-C, Hui P-C-L, Yeung K-W, Ng F-S-F (1998) Handling the assembly line balancing problem in the clothing industry using a genetic algorithm. Int J Clothing Sci Technol 10(1):21–37

Chau L (2012) The wasteful culture of Forever 21, H&M, and 'Fast Fashion'. http://www.usnews.com/opinion/blogs/economic-intelligence/2012/09/21/the-wasteful-culture-of-forever-21-hm-and-fast-fashion. Accessed 15 Feb 2016

Choi JI, Chung YJ, Kang DI, Lee KS, Lee JW (2012) Effect of radiation on disinfection and mechanical properties of Korean traditional paper, Hanji. Radiat Phys Chem 81(8):1051–1054. Accessed 4 Mar 2016

Cline E (2014) Where does discarded clothing go? http://www.theatlantic.com/business/archive/2014/07/where-does-discarded-clothing-go/374613/

Clinton (N.D.) Model behavior. http://www.sustainability.com/projects/business-model-innovation. Accessed 2 Mar 2016

Conca J (2015) Making climate change fashionable—the garment industry takes on global warming. http://www.forbes.com/sites/jamesconca/2015/12/03/making-climate-change-fashionable-the-garment-industry-takes-on-global-warming/#2af0bac778a2. Accessed 2 Mar 2016

Cooklin G (1991) Introduction to clothing manufacturing. Blackwell Science, Oxford 104

Council for Textile Recycling (N.D.) The facts about textile waste. http://www.weardonaterecycle.org/about/issue.html. Accessed 2 Mar 2016

Dagirmanjian J (N.D.) The future of 3D printing and sustainable fashion. https://www.purible.com/stories/3Dprinting. Accessed 16 Feb 2016

D'Arcy J-B (1986) Sheep and wool technology. NSW University Press, CRC Press

David (2015) 10 thins you should know about 3D knitting. http://comfedesigns.com/3d-knitting/. Accessed 13 Feb 2016

Deirdre S, Riach K (2011) Embracing ethical fields: constructing consumption in the margins. Eur J Mark 45(7–8):1051–1067

Ecofashiontalk (2012) Kotoba. http://www.ecofashiontalk.com/2012/09/kotoba/. Accessed 16 Feb 2016

Elks J (2014) Levi's Water<Less Jeans have saved 770 million liters so far. http://www.sustainablebrands.com/news_and_views/waste_not/jennifer_elks/levis_waterless_jeans_have_saved_770_million_liters_so_far. Accessed 22 Feb 2016

Emilia R (2015) Max Mara: A champion of Shima Seiki WHOLEGARMENT technology. http://www.knittingindustry.com/max-mara-a-champion-of-shima-seiki-wholegarment-technology/. Accessed 2 Mar 2016

Entrepreneur (N.D.) Packaging. https://www.entrepreneur.com/encyclopedia/packaging. Accessed 2 Mar 2016

Epstein-Reeves J (2012) Six reasons companies should embrace CSR. http://www.forbes.com/sites/csr/2012/02/21/six-reasons-companies-should-embrace-csr/#2421a1c34c03. Accessed 2 Mar 2016

Fashionbi (2013) Fast fashion market report. http://fashionbi.com/market/fast-fashion/all. Accessed 15 Feb 2016

Fetcher A (2012) Patagonia continues to lead base layer market with Fall'13 collection, 100 % Blue Sign® system approved fabric. http://www.patagoniaworks.com/Press/2014/6/24/patagoniacontinues-to-lead-baselayer-market-with-fall-13-collection-100-bluesign-approved-fabric. Accessed 14 Feb 2016

Fletcher K (2007) Slow fashion. http://www.theecologist.org/green_green_iving/clothing/269245/slow_fashion.html. Accessed 15 Feb 2016

Gardetti M, Muthu S (2015) Sustainable apparel? Is the innovation in the business model?—The case of IOU project. Text Clothing Sustain 1(1):1–9

Global Organic Textile Standard (N.D.) http://www.global-standard.org/about-us.html. Accessed 16 Feb 2016

Greenchoices (N.D.) Environmental impacts. http://www.greenchoices.org/green-living/clothes/more-sustainable-fabrics. Accessed 2 Mar 2016

Greenpeace (2010) The dirty secret behind jeans and bras. http://www.greenpeace.org/eastasia/news/stories/toxics/2010/textile-pollution-xintang-gurao/. Accessed 22 Feb 2016

Greenpeace.org. (2012) Toxic threads: the big fashion stitch up. Retrieved from http://www.greenpeace.org/international/Global/international/publications/toxics/Water%202012/ToxicThreads01.pdf. Accessed 4 Mar 2016

Gunther M (2013) Patagonia seeks more sustainable wool in Patagonia. https://www.greenbiz.com/blog/2013/02/11/patagonia-sustainable-wool. Accessed 22 Feb 2016

HBS Working Knowledge (2015) Stella McCartney: luxury and environmental sustainability can co-exist. Retrieved from http://www.forbes.com/sites/hbsworkingknowledge/2015/11/17/stella-mccartney-luxury-and-environmental-sustainability-can-co-exist/#3f075fb32fc0. Accessed 17 Feb 2016

Hecht J (2005) Beam: the race to make the laser. Oxford University Press, Oxford

Hethorn J, Ulasewicz C (2015) Sustainable fashion: what's next? A conversation about issues, practices and possibilities. Fairchild Books, New York

Hill A (2014) Eileen Fisher: timeless design 30 years later. http://www.marketplace.org/2014/12/29/business/eileen-fisher-timeless-design-30-years-later. Accessed 13 Feb 2016

Hunter (2015) Max Mara: A champion of Shima Seiki WHOLEGARMENT technology. http://www.knittingindustry.com/max-mara-a-champion-of-shima-seiki-wholegarment-technology/. Accessed 5 Mar 2016

IBISworld (2015) Global apparel manufacturing market research report. http://www.ibisworld.com/industry/global/global-apparel-manufacturing.html. Accessed 2 Mar 2016

Ifoam EU Group (2015) Climate benefits of organic cotton. http://www.ifoam-eu.org/en/news/2015/09/11/climate-benefits-organic-cotton. Accessed 22 Feb 2016

International Organization for Standardization (2006) ISO standards for life cycle assessment to promote sustainable development. http://www.iso.org/iso/home/news_index/news_archive/news.htm?refid=Ref1019. Accessed 5 Mar 2016

International Organization for Standardization (2015) ISO 14001:2015—Environmental management. http://www.iso.org/iso/iso14000. Accessed 2 Mar 2016

Isseymiyake (N.D.) ISSEY MIYAKE: Autumn winter 2016 Collection. http://www.isseymiyake.com/en/brands/issey_miyake.html. Accessed 12 Feb 2016

Jang YS, Amna T, Hassan MS, Kim HC, Kim JH, Baik SH, Khil MS (2015) Nanotitania/ mulberry fibers as novel textile with anti-yellowing and intrinsic antimicrobial properties. Ceram Int 41:6274–6280

Joy A, Sherry J-F, Venkatesh A, Wang J, Chan R (2012) Fast fashion, sustainability, and the ethical appeal of luxury brands. Fashion Theor 16(3):273–296

Jung J-Y, Lee J-H, Park C-G (2013) Effect of styrene–butadiene latex on the bond performance of macro synthetic fiber in micro jute/macro synthetic hybrid fiber-reinforced latex-modified cement-based composites. J Appl Polym Sci 127(5):3522–3529

Kadolph S-J (2007) Textiles, 10th edn. Prentice-Hall, Upper Saddle River, New Jersey, USA

Kant R (2012) Textile dyeing industry an environmental hazard. Natural Science 4(1):22–26

Khalil E (2015) Sustainable and ecological finishing technology for denim jeans. AASCIT Commun 2(5):159–163

Khan MR, Islam M (2015) Materials and manufacturing environmental sustainability evaluation of apparel product: knitted T-shirt case study. Textiles and Clothing Sustainability 1(1):1–12

Kim G (2016) ADIDAS, H&M among the world's most sustainable corporations. http:// fashionista.com/2016/01/most-sustainable-fashion-companies. Accessed 15 Feb 2016

King J (2012) Trash talking: textile recycling. https://waste-management-world.com/a/trash-talking-textile-recycling. Accessed 15 Feb 2016

Kinosian J (2015) Eileen Fisher's sustainable vision could make every day Earth Day. http://www. latimes.com/fashion/alltherage/la-ar-eileen-fishers-sustainable-vision-20150420-story.html. Accessed 13 Feb 2016

Kirchain R, Olivetti E, Miller T-R, Greene S (2015) Sustainable apparel materials. http://msl.mit. edu/publications/SustainableApparelMaterials.pdf. Accessed 18 Feb 2016

Lamicella L (2014) H&M named world's biggest user of organic cotton. https:// sourcingjournalonline.com/2013-organic-cotton-report-released-ll/. Accessed 13 Feb 2016

Lee (2016) Designer Kyung Eun Lee's portfolio. Ames: Lee, K

Levis (N.D.) Levi's® Water<Less™ Jeans—Finishing process. http://store.levi.com/waterless/. Accessed 2 Mar 2016

Levistrauss (N.D.) Water<Less™. http://www.levistrauss.com/sustainability/products/waterless/. Accessed 22 Feb 2016

Levy M, Weitz B-A (2008) Retailing management, 7th edn. McGraw-Hill Irwin, Boston

Lewis H, Gertsakis J, Grant T, Morelli N, Sweatman A (2001) Design environment: a global guide to designing greener goods. Sheffiled, Greenleaf, p16

Libolon (N.D.) RePET yarns. http://www.libolon.com/eco.php. Accessed 15 Feb 2016

Lights Z (N.D.) 5 Nontoxic alternatives to polyurethane. http://greenlivingideas.com/2014/07/28/ 5-nontoxic-alternatives-polyurethane/. Accessed 2 Mar 2016

Loeb W (2015) Zara leads in fast fashion. http://www.forbes.com/sites/walterloeb/2015/03/30/ zara-leads-in-fast-fashion/#54b17dfa61d7. Accessed 15 Feb 2016

MacDonald S (2012) U.S. Textile and Apparel Industries and Rural America. http://www.ers.usda. gov/topics/crops/cotton-wool/background/us-textile-and-apparel-industries-and-rural-america. aspx. Accessed 13 Feb 2016

Macysgreenliving (2013) Going green goes glamorous at Bloomingdale's. http:// macysgreenliving.com/sustainability-in-action/by-our-associates/going-green-goes-glamorous-at-bloomingdales/. Accessed 13 Feb 2016

Małgorzata Z, Marina M, Izabella K, Bogusław W (2003) The physical properties of the surface of apparel made from flax and polyester fibres. Int J Clothing Sci Technol 15(3/4):284–294

Marketwired (2014) Global apparel and footwear sales to Hit US$2 Trillion by 2018. http://www. marketwired.com/press-release/global-apparel-and-footwear-sales-to-hit-us2-trillion-by-2018-1886781.htm. Accessed 2 Mar 2016

McDonough W, Braungart M (2002) Remarking the way we make things: Cradle to Cradle. North Point Press, New York

McGregor L (2015) How target convinced its designers to embrace 3-D technology. https:// sourcingjournalonline.com/target-convinced-designers-embrace-3-d-technology-lm/. Accessed 4 Mar 2016

Nike (2013) NIKE partners with Bluesign technologies to scale sustainable textiles. http://news. nike.com/news/nike-partners-with-bluesign-technologies-to-scale-sustainable-textiles. Accessed 2 Mar 2016

Oecotextiles (2009) What does organic wool mean? https://oecotextiles.wordpress.com/2009/08/ 11/what-does-organic-wool-mean/. Accessed 15 Feb 2016

Oecotextiles (2010) Plastics—part 1. https://oecotextiles.wordpress.com/tag/recycle/. Accessed 13 Feb 2016

Oecotextiles (2012a) Eucalyptus fiber by any other name. https://oecotextiles.wordpress.com/tag/ regenerated-cellulose/. Accessed 2 Mar 2016

Oecotextiles (2012b) Textile printing and the environment. https://oecotextiles.wordpress.com/ 2012/01/27/textile-printing-and-the-environment/. Accessed 15 Feb 2016

Oecotextiles (2012c) What does "mercerized" cotton mean? https://oecotextiles.wordpress.com/ 2012/12/05/what-does-mercerized-cotton-mean/. Accessed 5 Feb 2016

Oecotextiles (2013) Fabric and your carbon footprint. https://oecotextiles.wordpress.com/tag/ greenhouse-gas/. Accessed 2 Mar 2016

Oecotextiles (2014) What does "eco friendly" vinyl mean? https://oecotextiles.wordpress.com/ category/chemicals/pvc/. Accessed 15 Feb 2016

Oecotextiles (N.D.) Textile industry poses environmental hazards. http://www.oecotextiles.com/ PDF/textile_industry_hazards.pdf Accessed 16 Feb 2016

OEKO-TEX® (N.D.a) Concept. https://www.oeko-tex.com/en/manufacturers/concept/concept_ start.html. Accessed 2 Mar 2016

OEKO-TEX® (N.D.b) OEKO-TEX® Standard 100. https://www.oeko-tex.com/en/Manufacturers/ concept/oeko_tex_standard_100/oeko_tex_standard_100.xhtml. Accessed 2 Mar 2016

Organiccotton (N.D.a) The risks of cotton farming. http://www.organiccotton.org/oc/Cotton-general/Impact-of-cotton/Risk-of-cotton-farming.php. Accessed 2 Mar 2016

Organiccotton (N.D.b) Agronomic practices. http://www.organiccotton.org/oc/Organic-cotton/ Agronomic-practice/Agronomic-practice.php. Accessed 2 Mar 2016

Optitex (2016) Optitex press kit. New York: Cove, S

Pacific Institute (2009) Energy implications of bottled water. http://www.pacinst.org/reports/ bottled_water/index.htm. Accessed 15 Feb 2016

Palamutcu S (2010) Electric energy consumption in the cotton textile processing stages. Energy 35 (7):2945–2952

Parsons School of Design (2011) Reap what you sew: Zero waste fashion at Parsons. http://blogs. newschool.edu/news/2011/02/zero-waste-fashion/#.VvtKAbR_dds. Accessed 6 Mar 2016

Patagonia (2015) Patagonia to cease purchasing wool from Ovis 21. http://www.patagonia.com/us/ patagonia.go?assetid=9895. Accessed 22 Feb 2016

Patagonia (N.D.a) Blue Sign® System. http://www.patagonia.com/us/patagonia.go?assetid=68401. Accessed 2 Mar 2016

Patagonia (N.D.b) Recycled polyester. http://www.patagonia.com/us/patagonia.go?assetid=2791. Accessed 15 Feb 2016

Patagonia (N.D.c) 1 % for the planet. http://www.patagonia.com/us/patagonia.go?assetid=81218. Accessed 2 Mar 2016

Perera P (2009) Greentailing. http://www.professionalmarketer.ca/Portals/0/Marketing% 20Canada/2015/Greentailing.pdf. Accessed 16 Feb 2016

Peta (N.D.) Environmental hazards of wool. http://www.peta.org/issues/animals-used-for-clothing/ wool-industry/wool-environmental-hazards/. Accessed 2 Mar 2016

Petrochemicals Europe (N.D.) Acrylic monomers. http://www.petrochemistry.eu/about-petrochemistry/products.html?filter_id=7. Accessed 2 Mar 2016

Phelan H (2012) The slow fashion brand: 10 brands that are doing it right. http://fashionista.com/ 2012/12/the-slow-fashion-movement-what-it-is-and-the-10-brands-that-are-doing-it-right. Accessed 13 Feb 2016

Puma (2010) Clever little bag by PUMA and Fuse project https://www.youtube.com/watch?v= vwRulz8hPKI. Accessed 16 Feb 2016

PVC.org (N.D.) PVC in consumer goods. http://www.pvc.org/en/p/pvc-in-consumer-goods. Accessed 15 Feb 2016

Radha K-V, Sridevi V, Kalaivani K (2009) Electrochemical oxidation for the treatment of textile industry wastewater. Bioresour Technol 100(2):987–990

Rosenboom S (2010) Fashion tries on zero waste design. http://www.nytimes.com/2010/08/15/fashion/15waste.html?_r=0. Accessed 11 Feb 2016

Roy Choudhury AK (2015) Development of eco-labels for sustainable textiles, Roadmap to Sustainable Textiles and Clothing, Text Sci Clothing Technol 137–173

Rydell R (2001) Nylon: The Story of a Fashion Revolution. Business History Review 75 (1):210–212

Samsung C&T (2016) Company press kit. Seoul, Korea: Shim, M. Santoni (N.D.) Seamless technology. http://www.santoni.com/seamless-technology.asp. Accessed 11 Feb 2016

SAEM Collection Ltd. (2016) Saem Collection Ltd.'s company report. Seoul: Kim, I-K

SecondHand 4 Business Ltd. (N.D.) Importance to recycle used clothing. http://www.secondhand4business.com/textile-recycling/importance-to-recycle-used-clothing/

Sellappa S, Prathyumnan S, Joseph S, Keyan K-S, Balachandar V (2010) Genotoxic effects in textile printing dye exposed workers by micronucleus assay. Asian Pac J Cancer Prev 11:919–922

Shima Seiki (N.D.a) Corporate social responsibility. http://www.shimaseiki.com/company/responsibility/. Accessed 11 Feb 2016

Shima Seiki (N.D.b) Wholegarment. http://www.shimaseiki.com/wholegarment/. Accessed 11 Feb 2016

Siegle L (2011) Why fast fashion is slow death for the planet. Zara's 200 designers come up with 40,000 designs each year, of which 12,000 are actually produced. http://www.theguardian.com/lifeandstyle/2011/may/08/fast-fashion-death-for-planet

Sivaramakrishnan CN (2009) Pollution in textile industry. Colourage 16(2):66–68

SPGprint (N.D.) Textile printing in a more sustainable way. http://www.spgprints.com/news/news/textile+printing+in+a+more+sustainable+way+?news_id=57. Accessed 11 Feb 2016

Steiner J, Steiner G (2009) Business, government, and society: managerial perspective. Text and case, 12th edn. McGraw-Hill Irwin, New York

Stern N-Z, Ander W-N (2008) Greentailing and other revolutions in retail: hot ideas that are grabbing customer's attention and raising profits. Wiley, Hoboken, New Jersey

Sweeny G (2015) It's the second dirtiest thing in the world—and you're wearing it. http://www.alternet.org/environment/its-second-dirtiest-thing-world-and-youre-wearing-it. Accessed 2 Mar 2016

Szaky T (2014) Finding recycling solutions for commercial packaging waste. http://www.packagingdigest.com/sustainable-packaging/finding-recycling-solutions-commercial-packaging-waste140528. Accessed 6 Mar 2016

Taylor N-F (2015) What is corporate social responsibility? http://www.businessnewsdaily.com/4679-corporate-social-responsibility.html. Accessed 2 Mar 2016

Teonline (2009) New approach of synthetic fibers industry, textile exchange. http://www.teonline.com/articles/2009/01/new-approach-of-synthetic-fibe.html. Accessed 15 Feb 2016

Textilelearner (N.D.) Seamless knitting technology: Benefits of seamless knitting technology. http://textilelearner.blogspot.com/2012/09/seamless-knitting-technology-benefits.html. Accessed 6 Mar 2016

Textile World (2015) Man-Made Fibers continue to grow. http://www.textileworld.com/textile-world/fiber-world/2015/02/man-made-fibers-continue-to-grow/. Accessed 5 Mar 2016

Theballofyarn (N.D.) What is Mercerized Cotton Yarn? http://www.theballofyarn.com/WordPress/2014/05/10/mercerized-cotton-yarn/. Accessed 5 Mar 2016

The New York Times (1991) Science watch; The nylon effect. http://www.nytimes.com/1991/02/26/science/science-watch-the-nylon-effect.html. Accessed 2 Mar 2016

Tokatli N (2008) Global sourcing insights from the clothing industry: the case of Zara, a fast fashion retailer. J Econ Geogr 8:21–38

US Environmental Protection Agency (N.D.a) Textile fabric printing. https://www3.epa.gov/ttnchie1/ap42/ch04/final/c4s11.pdf. Accessed 17 Feb 2016

US Environmental Protection Agency (N.D.b) Greenhouse gas equivalencies calculator. https://www.epa.gov/energy/greenhouse-gas-equivalencies-calculator. Accessed 4 Mar 2016

Van der Veldena NM, Kuuskb K, Köhlerc AR (2015) Life cycle assessment and eco-design of smart textiles: the importance of material selection demonstrated through e-textile product redesign. Materials & Design 84(5):313–324

Vinodh S, Rathod G (2010) Integration of ECQFD and LCA for sustainable product design. J Clean Prod 18(8):833–842

Zadro D (2014) Changing demographics are prompting retail shifts in marketing. http://www.business2community.com/marketing/changing-demographics-prompting-retail-shifts-marketing-0957892#e2O11A80DV3qstVX.97. Accessed 5 Mar 2016

Zhang Y, Liu X, Xiao R, Yuan Z (2015) Life cycle assessment of cotton T-shirts in China. Int J Life Cycle Assess 20(7):994–1004

Zhu O, Sarkis J, Geng Y (2005) Green supply chain management in China: pressures, practices and performance. Int J Oper Prod Manage 25(5):449–468

# Social Sustainability in Textile Industry

S. Grace Annapoorani

**Abstract** The textile and clothing related industry holds a remarkable position in the global merchandize trade across countries. Budding countries account for two-third of global exports in textiles and clothing. In the world textile market, USA and Europe import textiles and apparels from Asia which acts as a prime region. India's one of the oldest business is textiles, and it has a remarkably sturdy occurrence in the nationwide market and contributes about 14 % to industrial production, 4 % to gross domestic product (GDP), and 27 % to the country's foreign exchange inflows. It also provides direct employment to over 45 million people. Textile industry is retaining sustained growth by affording one of the most basic needs of people for developing quality of life, and it holds the importance. The ready-made garment (RMG) industry is one of the largest urban employers in India and is a key driver of the national economy. Over the past twenty years due to increase in the labour inputs the industry has changed from informal to formal factory based industry, which is highly dependent on labour inputs. This section of the book deals with social aspects of sustainability in various sectors of textile industry. The textile industry manufactures fabric from natural and man-made fibres. There are different stages of processing starting from sorting, roving, spinning, blending, and dyeing, and finally, the fabric is either weaved or knitted. Labour is vital to the sector's current competitiveness and long-term capability. Workers' skill levels, productivity, and motivation make the industry's ability to be a focus and retain the right quantity and quality of workers, domestic labour laws, and regulations, and workers' living conditions and costs in urban areas are all critical in the circumstances of a continuously changing economic environment. In South Asia and other emerging economies, where low-cost labour is extremely important for industry competitiveness, the clothing industry has been subject to different legal accusations of labour abuse, including long hours, forced overtime, and low wages. Because of these factors, there have been many state and non-state attempts to try to secure sound labour and other practices in the sector while

S. Grace Annapoorani (✉)
Department of Textiles and Apparel Design, Bharathiar University,
Coimbatore, Tamil Nadu, India
e-mail: gracetad11@gmail.com

© Springer Nature Singapore Pte Ltd. 2017
S.S. Muthu (ed.), *Sustainability in the Textile Industry*,
Textile Science and Clothing Technology, DOI 10.1007/978-981-10-2639-3_4

maintaining its international competitiveness. There are some major sustainability issues in each sector. There are risks in terms of worker abuse, types of wages, gender equality, child labour, etc. The sustainability issues differ from one sector to another. The issues might differ between work processes in a sector. Normally, the issues are of social, economic, and environmental sustainability. There are many ways to analyse the sustainability issues within a sector. It depends on market and suppliers, product demands, geographic location, technology and labour force demographics, and their skills.

**Keywords** Social sustainability · Gender equity · Wages · Work time · Safety hazards

## 1 Introduction

According to the World Trade Organization Report, textiles and clothing are measured as one of the major products in world trade. The textiles and apparel industry are confronting issues, for example cut-off points to common assets, global warming, sustainability issues, and other social and political patterns. Textiles and clothing commercial ventures are considered among the most ecologically harming and untimely item substitution against environment. The textile industry is one of the greatest users of water. However, this industry has been becoming gigantic as are the ecological issues connected with it.

With developing alertness, sustainability increased main impetus in the textile and apparel industry in 1994, and today, it is turning out to be considerably more imperative. There are a few sustainability issues that particularly identify with sourcing, creation, fabricating, bundling, marketing, and utilization. There are two alternatives in managing sustainability issues: (1) it is possible that we can disregard them or (2) we can turn into the specialists of progress.

However, sustainability, much the same as some other significant origination, is a troublesome one to characterize in textiles and apparel.

## 2 Social Sustainability—Meaning

Social sustainability is the capacity of a group of people to build up configurations to meet the requirements of its existing constituent and also to support the ability of future generations to sustain a healthy community.

The method of approaching sustainability and its development is also known as social sustainability. The term social sustainability has less awareness among people when compared to economic and environmental sustainability.

The common meaning of social sustainability is the capacity of a societal system, like a country, to perform at a distinct level of social well-being indefinitely.

That level ought to be characterized in connection with the objective of Homo sapiens, which is to improve personal satisfaction for those living and their generations. Social manageability includes human rights, work rights, and corporate administration. Social assets incorporate thoughts as wide as different societies and fundamental human rights.

According to Brundtland Commission (1987), sustainability is frequently characterized as "addressing the necessities of today without trading off the ability of future eras to meet their individual prerequisites".

Sustainability triad contains environmental sustainability, economic sustainability, and social sustainability. It is otherwise called as interchange of economic, social, and environmental factors of advancement. The communication of these elements under enhancement frames a triad for feasible improvement.

The idea of "social sustainability" comprises of social value, liveability, well-being value, group improvement, social capital, social bolster, human rights, work rights, place making, social obligation, social equity, social skill, group versatility, and human adjustment.

Sustainability is the effective meeting of present social, economic, and ecological requirements without trading off the capacity of future era to address their own issues, got from the most widely recognized meaning of supportability.

Sustainability is coordinating human prosperity with accepted reliability. Social obligation incorporates sustainability with specific issues such as asset utilization, contamination, purchaser prosperity, human rights, well-being and security, item moderateness, and quality. Social obligation is a cognisant push to keep up and maintain human well-being and prosperity while being ecologically capable.

Sustainable improvement is a demanding expression, and only few individuals concur on what it implies. Individuals can take the term and "reinvent" it thinking of one's own particular needs. It is a thought that constantly drives us to change objectives and priority since it is an open procedure and all things considered, it cannot be come to authoritatively. The critical objective of this improvement model is to raise the personal satisfaction by long-haul amplification of the gainful capability of biological communities, through the proper innovations for this reason.

Achieving sustainability is the objective of economical improvement. "Sustainability" has a few implications and is frequently connected with social, environment and economic factors. Sustainability is the harmony between three components: economy, environment, and social value.

Sustainable advancement is another idea, as well as another state of mind, and this requests us to take a gander at things in an unexpected way. It is a thought of the world profoundly not the same as the one that principles our present thinking and incorporates fulfilling fundamental human needs, for example equity, flexibility, and poise. It is the vision through which we can manufacture a method for being. Sustainability is at the individual level, as the appraisal of every human conduct with the vision of reformulating those that negate the advancement of a reasonable future.

# 3 Working Environment and Hazards in Textile Industry

Dating back several centuries, the textile sector is one of the oldest industries in a country's economy. Till present, textile sector is one of the prime providers to country's exports with approximately 11 % of total exports. The textiles industry is one the largest labour-oriented industry.

The textile industry is chiefly concerned with the design and production of yarn, cloth, clothing, and their distribution. The raw material may be natural or synthetic using products of the chemical industry. There are numerous safety and health problems associated with the textile industry.

## 3.1 Major Hazards in the Textile Industry

The hazards happening in the textile industries are mechanical hazards, physical hazards, chemical hazards, ergonomic hazards, and physiological hazards

- Mechanical hazards,
- Chemical hazards,
- Biological agent hazards,
- Ergonomic hazards, and
- Psychosocial Hazards.

### 3.1.1 Mechanical Hazards—Cotton Dust

The workers working in the processing and spinning of cotton are exposed to large amounts of cotton dust. Cotton dust implies "dust present is noticeable all around amid taking care of or handling of cotton". This dust contains a blend of segments which may incorporate ground-level plant matter, cotton fibre lints, microscopic organisms, parasites soil, and pesticides. Manufacturing processes using new or waste cotton fibres or cotton fibre by-products from textile mills also produce cotton dust. They are also in contact with small particles of dust and pesticides. Continuous inhaling of cotton dust and other particles leads to respiratory disorders among the textile workers. Due to over exposure to cotton dust, the disease called byssinosis, also known as brown lung, is found among people working in the textile industry. The common symptoms of this disease include chest congestion, coughing, wheezing, and breathing suffocation.

OSHA—the occupational safety and health administration—made it mandatory for textile industry workers to safeguard their employers from overexposure to cotton dust and its evil effects. The OSHA figured out certain guidelines which are related to all private employers in the US textile industry.

A study identified with textile units in India was focussed in the Year 2007 which found that aspiratory capacity in textile labourers expanded altogether with introduction to cotton dust over a drawn-out stretch of time. Another study identified with textile units in Mumbai, India, demonstrated an 11–33 % frequency of constant bronchitis among labourers working in textile industry. Another study uncovered an expansion in the rate with an expansion in presentation to cotton dust.

Researchers have additionally stated that intense respiratory sicknesses are more normal among the kids working in floor covering weaving units in Jaipur. The commonness of respiratory sicknesses among child textile labourers was 26.4. Experts trust this is by virtue of high inhaling of cotton dust.

The occupational safety and health administration have set out a Cotton Dust Standard with a perspective to diminishing the exposure of the workers to cotton and shielding them from the danger of byssinosis. It has set up permissible exposure limits (PELs) for cotton dust for various operations in the textile organization. This has cut down the rate of event of byssinosis altogether. Diverse states may embrace distinctive guidelines for word-related security and well-being; notwithstanding, in those states where there are no gauges settled by the state, the federal norms are acknowledged.

For an eight-hour day, the OSHA Cotton standard has been chosen, at 200 mg of cotton dust per cubic metre of air if there should be an occurrence of yarn assembling, 500 mg in the event of material waste houses, 750 mg if there should be an occurrence of weaving operations, and 1000 mg if there should be an occurrence of for waste reusing. Textile industries are requested to calculate approximately the amount of respirable cotton clean once in 6 months, or at whatever point, there is any change that may prompt an adjustment in the level of dust. In the event that the level of dust in the air is higher than that according to OSHA rules, the administration ought to take measures to lessen the same. According to these rules, the employers must advise the workers in composing of the dust level present in the air and the strides that the administration wants to take for its decrease. On the off chance that the dust level cannot be diminished, it is the obligation of the administration to give respirators to the labourers. The OSHA Cotton Dust Standard was altered in the year 2000, which exempted a technique for washing cotton from the principle.

### 3.1.2 Chemical Hazards

Labourers in the textile industry are likewise exposed to some of chemicals, particularly those working in the section of dyeing, printing, and finishing. Chemicals taking into account benzidine, optical brighteners, solvents and fixatives, wrinkle resistance discharging formaldehyde, fire retardants that incorporate organophosphorus and organobromine mixes, and antimicrobial agents are utilized as a part of textile operations.

Textiles fibres, reactive dyes, synthetic fibres, and formaldehyde can be found as respiratory and skin sensitizers in the textile industry. The textile industry has been assessed as an area with an expanded cancer-causing hazard. A few studies have demonstrated an expanded danger of nasal, laryngeal, and bladder tumour in ladies.

There is no confirmation to recommend that most of the dyestuffs presently utilized as a part of dyeing and finishing are unsafe to human well-being, but it depends upon the level the labourers are exposed to the hazardous chemicals. Reactive dye is the most common hazard for respiratory issues because of the inhalation of dye particles. The respiratory sensitization and indications incorporate tingling, watery eyes, sniffling, and manifestations of asthma, for example cough and wheezing.

Many researches have indicated connection between introduction to formaldehyde and nasal and lung cancer and additionally to brain cancer and leukaemia, which can be lethal. Introduction to formaldehyde could prompt respiratory trouble and skin inflammation.

A research was conducted in USA that the people working in textile industry are prone more to mouth and throat cancer. Another study reveals that the textile workers higher chances of stomach and oesophageal cancer. Above all, there are higher chances of colorectal growth, thyroid disease, testicular malignancy, and nasal tumour.

As per the research conducted in Jodhpur "tie and dye" units out of 1300 workers, it is observed that nearly 100 workers are subjected to skin-related diseases. The main reason is because of Red RC base and naphthol.

### 3.1.3 Biological Hazards

Certain biological agents such as anthrax, clostridium tetani (tetanus), and coxiella burnetti affect the textile workers, when they are involved in the activity such as carding and roving which can cause allergies and breathing disorders.

### 3.1.4 Ergonomic Hazards

This session tells about the accumulation of workers, improper condition of the machine, ergonomic problem faced by the worker, dust problems, poor lighting, and ventilation and unaware of personal protective equipment. In textile industry, ergonomic issues are seen in a many of the units. The vast majority of these units have a workplace that is hazardous and undesirable for the labourers. The labourers in these units confront a few issues, for example no great furniture, despicable ventilation and lighting, and absence of proficient well-being measures if there should be an occurrence of crises. The labourers in such units are at danger for creating different occupational diseases. Carpal passage disorder, epicondylitis, lower arm tendinitis, lower back agony, neck torment, bicapital tendinitis, shoulder torment, and osteoarthritis of the knees are some of the musculoskeletal disarranges that have been seen among the labourers because of poor ergonomic conditions. These problems are more normal in new countries when contrasted with developed ones.

The stools and the tables used for different performance such as cutting and ironing were found to have a big difference in the heights. This prompted the specialists sitting in an uncomfortable position for entire work days. The stools were not cushioned in the majority of the units, prompting expanded uneasiness with respect to the specialists. The stools did not have a backrest, as a consequence of which the workers did not get enough backing to the back. In the greater part of the units, there was a low-level lighting prompting eye strain. The humidity level increases in the textile units because of continuous ironing which adds to the labourers' distress.

Musculoskeletal Disorders

The most widely recognized work-related health issues in Europe are musculoskeletal disorders (MSDs). In the textile sector, workers lifting the heavy goods, holding them, keeping it down, and pulling or pushing the goods were found to be the greatest reason for muscular disorders. Normally, manual handling of goods were found to be one of the gradual reasons behind causing deterioration of the musculoskeletal system, such as lower back pain, neck pain, and shoulder pain.

The hazard components for MSDs in the textile sector include the posture of the workers in different stages of textile sector. The MSDs are caused by continuous work, lifting heavy weight, and doing job without appropriate procedures, as quotes Tiwari (2012).

### 3.1.5 Psychosocial Problems in the Textiles Industry

Anxiety related to work stress has been characterized when the requests of the workplace surpass the labourers' capacity to adapt or control them. Work-related anxiety might be an issue in a few territories of the textile industry, being related, for instance, with dreary and quick paced work and where the labourer has no impact on how the job is finished. Padmini (2012). Education is the fundamental right that helps the growth of nation. The education helps the workers to get knowledge about medical rights and legal and social behaviour. The people are uneducated, and most of them do not know OHS at workplace. The company is unaware of OHS and also lacks in training, housekeeping, accident prevention, hospital facility, safety signs etc quotes Malik (2010).

## 3.2 Other Common Hazards

### 3.2.1 Noise Hazards

Textile industry is complained to be the highest noise polluting industry. Due to over subjection to high level of noise, there are chances for ear drum damage and hearing loss. Different issues such as weariness, absenteeism, irritation, uneasiness, decrease in effectiveness, changes in pulse rate and blood pressure, and insomnia are also stated because of continuous exposure to noise. Non-maintenance of machine and its parts are one of the reasons behind the 3 noise pollution. Despite the fact that it causes genuine health issues, noise pollution is regularly disregarded by textile units since its belongings are not promptly unmistakable and there is a non-appearance of torment. In industries, noise is a big problem that affects the human peace and increases the stress. The main cause of noise problem in the weaving and spinning industry is due to the poor design, overload, and old machineries. To control the noise level in the company premises and outside the company, necessary action of noise regulation must be adopted. To maintain the quality and production, the health of worker is essential. The most important hazard in occupation is noise. To maintain the quality and production, the health of the worker is essential says Ahmad et al. (2001).

### 3.2.2 Hazards from Fire

The fire accidents in textile industry are very common, and it sometimes creates a huge loss to the industry. Among textile industries in the world, Bangladesh was found to be the worst in terms of fire safety. During 2006–2009, about 414 textile workers died due to industry fire in the 213 textile industry. In 2010, about 79 workers died due to factory fires (Clean Clothes Campaign 2012).

In many textile industries, buildings are constructed in such away which is not fittingly intended for high electrical circuits expected to run an industry. The vast majority of these structures are found with poor electrical circuits. Additionally, poor electrical wiring, dubious and precarious force supply that makes short circuits, unprotected electrical outlets, dust and combustible materials encompassing electric outlets, and overall absence of mindfulness are the fundamental drivers of flame mishap. Numerous production lines run throughout the day and throughout the night without shutdown all through the entire week or considerably more keeping in mind the end goal to take care of the supply demand.

### 3.2.3 Accident in the Textile Industry

There are numerous perils in the textile industry that can make damage labourers, from transportation in the work environment, unsafe extensive work gear, and plant, to the danger of slips from a wet workplace. Workers report that their trespass is being blocked by the movement of heavy machinery, vans, and truck which are a noteworthy reason for accidents. There additionally exist the dangers of flame and blasts.

### 3.2.4  Poor Building Construction

The major hazard was found among textile industry building collapse due to poor building construction. The principle reason for building breakdown is unapproved and critical structure of the building. Numerous industries unlawfully develop their building without having legitimate consent. The basement of the building does not sustain additional burden and subsequently the building breakdown.

The next reason for building collapse is before constructing a building, there was no legitimate soil test and site examination done. Many buildings were constructed in area where water body is found. Without proper permission from the land authorities, the heavy structure building is constructed which cannot withstand any heavy workload and pressures.

## 3.3  Sustainability in Working Environment of the Textile Industry

### 3.3.1  Responsibility of the Organization

The administration or the organization has to routinely check and report the national laws and regulations concerning working environment security. The administration ought to then build up a convention through which to actualize these laws. It might likewise be important to consider the necessities of specific purchasers, who may have implicit rules that incorporate parts of H&S, corporate social obligation, and natural obligation.

### 3.3.2  Basic Necessities

Workers should be offered access to safe drinking water and additionally hygienic place for them to have their food. This place should be away from production area. The industry should provide adequate number of toilets for the workers. This is a legal necessity.

### 3.3.3  Documentation Maintenance

Records of wounds during working time ought to be made for arranging future safety. The administration ought to build up an agenda of measures and activities that should be led month to month to guarantee that welfare rules are being taken after and to research incidents about accidents. This should be possible by means of industrial facility visits and searching for potential risks in the work environment, checking the mishap and comfort records, and approaching the representatives for criticism on health and safety issues. Administration ought to likewise have a support plan to decrease accidents.

### 3.3.4  Symbols and Signs

Images and signs are an essential method for illuminating and helping staff regarding safety. Sufficient number of fire extinguishers should be made accessible and signs to be placed at important places. The workers should be educated about their use and importance, and it should be easily accessible. Fire alarms and danger lights should be made visible to all. Chemicals that are hazardous should be labelled properly, and the workers should be educated about the chemicals.

### 3.3.5  First Aid

At least one individual from workers should be trained on medical usage of first-aid kit in case of all emergency treatment necessities amid their day of work. A code of conduct is additionally required to guarantee that each labourer knows who the medical aid individual is so that they can get in touch with them rapidly in a crisis. At least one first-aid box should be made accessible to the labourers in their working area. The first-aid kit should contain the following:

- Sterile cotton,
- Band aid,
- Sterile gauze,
- Scissors,
- Ointment,
- Liquid antiseptic, and
- Pain killers.

The emergency treatment box should be legitimately maintained up by a trained person and checked consistently. A report book should be maintained, and each accidents or injuries should be recorded. This facilitates the organization to further avoid accidents or keep them in control.

## 4  Wages and Work Time

Wages and working time directly affect the lives of specialists and the intensity of organizations. A key part of the International Labour Work—ILO's effort is the improvement of international labour standards that help governments in setting up national legislation to direct wages and working hours and give businesses' and specialists' delegates with a strong legitimate structure for aggregate bartering and different types of arrangement.

## 4.1 Wages Regulation

Wage regulation and compensation setting have been a basic part of the ILO order from its beginning. The preamble of the ILO Constitution of 1919 and the Declaration of Philadelphia of 1944 call for "approaches concerning wages and income, hours, and different states of work figured to guarantee a simple share of the products of advancement to all and a base living pay to all utilized and needing such security". The privilege to "simple and ideal compensation" that guarantees a presence deserving of human poise was perceived as a major human right in The Universal Declaration of Human Rights in 1948. The ILO Tripartite Declaration of Principles concerning Multinational Enterprises and Social Policy, as changed in 2006, alludes to the requirement for multinational undertakings (MNEs) to regard wage levels and "give the most ideal wages, advantages, and states of work, inside the structure of government arrangements". In 2008, the ILO Declaration on Social Justice for a Fair Globalization, consistently embraced by the International Labour Conference (ILC), expressed that "accomplishing an enhanced and reasonable result for all has turned out to be significantly more vital" and reviewed the ILO's commitment to advance the targets of "full work and the raising of ways of life, a base living compensation".

The textile sector is of significant importance to the Indian economy. Not just contributes the business generously to India's fare gaining, it is evaluated that one out of each six families in the nation relies upon this segment, either straightfor-wardly or in a roundabout way, for its employment.

The following are some of the issues related to wages in the textile industry.

- **Minimum Wages**: Workers cannot manage their daily life expenses with the insufficient income they earn from the industry.
- **High** Work **Pressure**: Due to rise in orders, the targets are fixed on hourly basis. Hourly targets are set way higher than what a healthy worker of average skill can produce.
- **Verbal Abuse**: when the target is not attained, the workers are verbally abused by their supervisors or by their higher authorities.
- **Overtime** is frequently not wilful by the workers. They are compelled to stay longer to finish targets. The legally set double normal hourly wage compensation for additional time is generally not paid.
- **Insecurity of Job**: Workers experience great employment unreliability. Workers feel steady risk of being rejected or dismissed. Negligible mistakes in work, non-completion of targets, reporting late to work even by a few minutes, and replying back when they are yelled at are all used as they are all utilized as ground for rejection. Furthermore, more than half of the workers will not sign any sort of agreement when they started working for a factory.
- **Rules and Regulations of the factory**: Most labours do not know about sets of principles. They are not given any kind of basic education about the rules or code of conduct to be followed during working time.

- **Labour union**: Many factories do not have labour union. In case of any problem, the labourers confront the issue to the administration and reporting a problem quite often works out counterproductive for the worker. Even if the workers are dynamic for a union, they are degraded and demotivated by directors and administration.

The pay board is an association under Ministry of Labour and Employment who chooses the compensation scale from government's point of view. Extra minutes pay and sponsorships have been incorporated to the normal compensation of the labourers in a few industrial facilities; however, the workers do not get it in time. Much of the time, additional time hours are not recorded legitimately.

The workers and labour unions asserted that the latest wage finalized by the pay board is not sufficiently adequate to meet the present high expenses of living and expanding pattern of expansion. Workers regularly show disagreement against minimum wages, and call strikes which prompt tremendous measure of misfortune to the industry management, put danger to country's economy.

Working excessively overtime is another issue. Labourers frequently do additional time because to get additional pay to take care of high living expenses. In most cases, the labourers are forced to work additional time because to attain the target. Normally, it is common that the labours have to work eight hours per day and six days per week. Totally, their workload is 48 h per week. But in many textile industries, they are compelled to work for 60 h rather than 48 h.

Kakuli and Risberg (2012) state that workers in Bangladesh textile industry work normally 76 h per week which is beyond the constraint. The labour act 2006 likewise expresses that a labourer ought to have been paid within 7 working days after completion of work. Shockingly, a large portion of the factories essentially do not tail this. In common, around 50 % benefit of the business should reach to labourer's wage which is commonly followed around the world. Many textile industries in the world do not share the exact percentage of profit for the welfare of workers.

The wage board considers several issues to prepare for a decent minimum wage level such as basic living standards, costs of living, production cost, cost of the produced goods, inflation rate, job types, business capacity, and socio-economic condition (Yunus and Yamagata 2012). Several controversies appear with the living cost estimation and with other issues from the government and different non-government agencies. Thus, the minimum wage calculation does not often reflect the reality.

The wage board takes into account a few issues to get ready for a fair, the lowest pay permitted by law level, i.e. to meet the day-to-day expenses. Living expenditure, creation cost, expense of the delivered merchandize, inflation rate, work sorts, business limit, and financial condition, and so forth report Yunus and Yamagata (2012). A few discussions show up with the living cost estimation and with different issues from the administration and distinctive non-government offices. Thus, the minimum wage computation does not frequently replicate the reality.

## *4.2   Overtime*

As indicated by the financial times, a large number of textile industries rush for the last-minute operation of work which forces the labourers to work overtime. This happens because of poor administration by the organizations requesting the materials. Very late outline, design, material selection, and even colour coding change, anticipate, and frequently make more work in the textile industry without a comparable augmentation to the conveyance due date. To meet the due date, requests of client's workers in the factory should regularly work broad extra time. An extensive variety of additional time, in any case, builds the dangers of specialist harm and diminishes profitability. Absence of critical, for example, medical aid packs, fire quenchers, and alerts was noted in the vast majority of the units. This puts the labourers under extraordinary danger in times of a crisis. Defensive types of gear-like metallic gloves were not given to the workers in a few units for insurance against conceivable occasions, accidents, and wounds.

## 5   Gender Equality

Today, in the society, women are playing a different role and frequently they are able to handle two or more undertakings at the same time. They are inclined to experience from work-related diseases, which are interrelated by social, physical, and physiological issues. About, 1 out of 300 females is experiencing from some occupation-related disease.

The social imperceptibility of womens' paid work is utilized to legitimize paying them lower wages than men. Women as beings are additionally rendered defenseless against tolerating low wages since they themselves see their paid work as less noteworthy than their essential assignment of home production. "Their misuse is undetectable behind a belief system that veils the way that they work by any means—their work seems inessential" says Juliet Mitchell in Woman's Estate.

## *5.1   Workplace Gender Equality Act 2012*

The equal opportunity for Women in the Workplace Act 1999 was replaced by the Workplace Gender Equality Act 2012. The enactment requires non-public sector employers with more than hundred workers to account for the agency yearly. In the case if the quantity of workers falls beneath 100, it must keep on reporting unless its number of representatives falls underneath 80. The Workplace Gender Equality Agency is controlling the act.

## 5.2   Safety and Health of Women in the Textile Sector

Since the exposure of women to dangers is contrasted with men, occupational safety and health (OSH) should be handled in a gender-sensitive way. Sexual orientation delicate intercessions ought to be participatory, including the specialists concerned, and take into account an examination of the genuine work circumstance. There must be an obligation from administration to take health and sexual orientation issues sincerely, and no suppositions ought to be made about who is at danger from what perils.

Pregnant or nursing labourers are ensured by a particular mandate, Council Directive 92/85/EEC, which sets up least norms and is transposed into each member states. The order obliges employers to complete a particular risk evaluation to the labourers secured and highlights specifically.

Bangladesh textile industries have given a colossal chance to women workers. As per the article of Nidhi (2009), among 1.8 million of textile workers of 3480 units, the women labourers were about 1.5 million. It demonstrates around 80 % of women textile workers.

The absence of training and absence of work ability draw in the women to work in textile units even with low wages. It is found that majority of women workers were doing household work or agriculture before getting employment in the textile units. This created a greater chance for the textile unit managers to take advantage over the situation.

It is found from many literatures that most of the women workers were young and unmarried. Kabeer (2004) guaranteed that around 40–50 % of textile women workers were married and expecting mothers. The main considerations that drive them to work in the textile industry are neediness, family strife, and separation Nidhi (2009). Working in the textile-based companies gives the women labourer's better monetary benefits even with a low wages. There are a few imbalances at work. Ladies are frequently utilized in occupations with less or non-specialized ability contrast with men. At times, female labourers are paid lower than that of the man with comparative employment. Inappropriate behaviour with women workers are another basic offence. Numerous female labourers have badgering encounters amid getting compensation or working at a night shift. In the majority of the cases, the manager or senior administrators harass them. Many women workers do not lodge complaint against them fearing that they might lose their job. Inexperience in job and non-eligibility force them to keep quiet against the sexual brutality. Now and then, during salary instalment time, female labourers are forced to pay for any fund failing to which results in physical torment.

# 6   Child Labour

Child labour is illegal by law in many nations. It is an offensive act to make the children to work hazardous, overwhelming, or heavy tasks. The UN Convention on Child Rights states that all work done by youngsters less than 15 years old and all unsafe work done by kids less than 18 years old is unlawful. But there are an expected 168 million to 200 million child workers working in many parts of the world. Though child labour has been abolished in many countries, it is still found to be around 11 % of world child population, as indicated by figures from the International Labour Organization (ILO).

Throughout the years, there has been a movement in the textile units that utilize child work. Priorly, child work was broadly utilized in export units, representing almost 60 % child work utilized in Tirupur. Due to the compulsion from the global buyers as a social consistence it has rendered a substantial number of textile units free from child work over a time frame.

Child work is additionally found in textile export units. It is found that one of the notorious industries that flourish with child labours over the many nations is Zari work—sequin work. It is a delicate embroidery that has turned out to be hugely famous in American and European design stores. Sweatshop proprietors like to utilize kids on the grounds that their meagre, small fingers can work speedier on many-sided ethnic plans. Times online states that it is assessed that 100,000 youngsters work for over 14 h a day in the unlawful sweatshops in and around Delhi. This makes the child fingers and hands gravely harmed and their visual perception frail from extend periods of time of monotonous work in darkrooms. Their physical growth is regularly hindered in uncomfortable sitting positions, slouched positions at the bamboo surrounded workstations. It is also found there are no fixed hours of work for them and there is no labour union to fight for them.

The textile business likewise confronts a continuous issue with the utilization of child work, comprehensively characterized by the International Labour Organization as work performed by anybody between the ages of 5 and 18 that abuses the child's human rights, safety, and educational opportunities. The ILO does, in any case, set criteria for non-exploitative youngster work. Regardless of universal shock and changes to inward arrangements by expansive organizations that puts pressure on textile units to keep children out of the creation process, Fashion Mag reports that an expected 168 million kids still work in the industry starting 2012.

The status of child labour in India is pathetic. It is found that about 35 % of them are working in unclean and dangerous jobs for their livelihood in Tirupur district. It is mandatory to abolish child labour in any industry, but research statistics state that in Tirupur textile units, more than 40,000 children are working under unsafe conditions.

## 6.1 Recommendations to Eliminate Child Labour

As child labour laws are subject to change, employers and managers need to keep up-to-date informations.

- Employers must rigorously check the identity papers of any prospective recruiter who appears to be below the age of 20 years.
- Young workers aged 15–17 years should work only in tasks appropriate to their age and stage of development, and not in activities which may damage their health or well-being. Employers may put in place a system of paid apprenticeship/vocational training for young people, with appropriate accreditation from the government.
- Employers should rigorously screen identity documentation for young workers already employed, to seek to identify any cases of falsified records. Should underage children have been mistakenly recruited, every assistance should be provided to ensure that they are removed from employment under the best conditions and not further disadvantaged; e.g., they should retain right to earn employee insurance benefits and should be assisted to find appropriate educational, training, or other opportunities appropriate to their age.

## 7 Labour Rights

### International Labour Standards

ILO from 1919 has kept up and built up an arrangement of international labour standards went for advancing open doors for women and men to acquire respectable and beneficial work, in states of flexibility, value, security, and nobility. International labour standards are stated to be a fundamental part in the universal structure in today's world economy for guaranteeing that the development of the worldwide economy gives benefits to all.

It is stated in the World Commission on the Social Dimension of Globalization, (ILO-1-2004) "The *principles of the worldwide economy ought to be gone for enhancing the rights, vocations, security, and chances of individuals, families and groups far and wide*".

The international labour standards have been framed for many associations. They are as follows.

## 7.1 Standards on Freedom of Association

The guideline of opportunity of affiliation is at the centre of the ILO's qualities: it is cherished in the ILO Constitution (1919), the ILO Declaration of Philadelphia

(1944), and the ILO Declaration on Fundamental Principles and Rights at Work (1998). It is likewise a privilege broadcasted in the Universal Declaration of Human Rights (1948). The privilege to arrange and shape employers and employees' associations is essential for sound aggregate haggling and social dialog. ILO norms, in conjunction with the work of the Committee on Freedom of Association and other supervisory systems, prepare for determining these challenges and guaranteeing that this key human right is regarded world over.

## 7.2 Standards on Collective Bargaining

Flexibility of affiliation guarantees that workers and employers can partner to effectively arrange work relations. Consolidated with solid opportunity of affiliation, sound aggregate haggling rehearses guarantee that businesses and labourers have an equivalent voice in arrangements and that the result will be reasonable and fair. Aggregate haggling permits both sides to arrange a reasonable work relationship and averts expensive work debate. For sure, some examination has demonstrated that nations with very planned aggregate haggling have a tendency to have less disparity in wages, lower and less tenacious unemployment, and less and shorter strikes than nations where aggregate bartering is less settled. ILO norms advance aggregate bartering and guarantee that great work relations advantage everybody.

## 7.3 Standards on Forced Labour

ILO gauges demonstrate that 20.9 million individuals around the globe are still subjected to it. Of the aggregate number of casualties of constrained work, 18.7 million (90 for every penny) are misused in the private economy, by people or undertakings, and the staying 2.2 million (10 for each penny) are in state-forced types of constrained work. Among those abused by private people or endeavours, 4.5 million (22 for each penny) are casualties of constrained sexual misuse and 14.2 million (68 for each penny) are casualties of constrained work abuse. Constrained work in the private economy produces US$ 150 billion in illicit benefits for every year: 66 % of the assessed aggregate (or US$ 99 billion) originates from business sexual abuse, while another US$ 51 billion results from constrained financial misuse, including household work, farming, and other monetary exercises.

In some parts of Africa, slaves are still found, while forced work as coercive and tricky enlistment is available in numerous nations of Latin America, and in some other countries. In various nations, domestic workers are caught in circumstances of constrained work, and much of the time they are controlled from leaving the owners home through dangers or savagery. Fortified work holds on in South Asia where a huge number of men, ladies, and youngsters are fixing to their work through an endless loop of obligation. In Europe and North America, an expanding number of ladies and youngsters are casualties of trafficking for work and sexual misuse. Trafficking in persons has been the subject of developing universal consideration as of late. At last, constrained work is still forced by the state for the motivations behind monetary improvement or as a discipline, including for communicating political perspectives.

For some legislatures around the globe, the disposal of constrained work remains an imperative test for the twenty-first century. Not just is constrained work a genuine infringement of a major human right, it is a main source of neediness and an obstacle to monetary improvement. ILO measures on constrained work and the remarks of the supervisory bodies, in mix with experience from specialized help and collaboration, have given imperative direction to part states to build up an exhaustive reaction to constrained work.

## 7.4 Standards on Child Labour

Child labour is an infringement of crucial human rights and has been appeared to impede children's improvement, possibly prompting deep-rooted physical or mental harm. Proof focuses to a solid connection between family neediness and tyke work, and kid work sustains destitution crosswise over eras by keeping offspring of the poor out of school and restricting their prospects for upward social versatility. This bringing down of human capital has been connected to moderate financial development and social advancement. A late ILO study has demonstrated that killing youngster work experiencing significant change and creating economies could produce monetary advantages almost seven times more noteworthy than the expenses, for the most part connected with interest in better educating and social administrations. ILO gauges on tyke work are essential global legitimate apparatuses for battling this issue.

## 7.5 Standards on Equality of Opportunity and Treatment

A huge number of ladies and men around the globe are denied access to employments and get low wages or are limited to specific occupations basically on the premise of their sex, skin shading, ethnicity, or convictions, without respect to their abilities and aptitudes. In various created nations, for instance, ladies' specialists gain up to 25 % not as much as male associates performing rise to work. Opportunity from segregation is a principal human right and is fundamental for both labourers to pick their livelihood uninhibitedly, to build up their capability to the full and to harvest financial prizes on the premise of legitimacy. Conveying balance to the work environment has critical monetary advantages, as well. Managers who rehearse correspondence have admittance to a bigger and more differing workforce. Labourers who appreciate balance have more noteworthy access to frequently get higher wages and enhance the general nature of the workforce. The benefits of a globalized economy are all more genuinely dispersed in a general public with correspondence, prompting more noteworthy social security and more extensive open backing for further financial advancement. ILO principles on fairness give apparatuses to take out separation in all parts of the working environment and in the public arena all in all. They additionally give the premise, whereupon sexual orientation mainstreaming methodologies can be connected in the field of work.

## 7.6 Standards on Employment Policy

For the vast majority, the way to getting away destitution implies having a vocation. Perceiving that creating work principles without tending to livelihood would be silly, the ILO devotes an extensive piece of its project to making more prominent open doors for ladies and men to secure better than average job and salary. To achieve this objective, it advances worldwide models on occupation approach which, together with specialized participation projects, are gone for accomplishing full, profitable, and uninhibitedly picked work. No single strategy can be endorsed to accomplish this goal. Each nation, whether creating, created, or on the move, needs to devise its own particular strategies to achieve full business. ILO gauges on occupational strategy give apparatuses to planning and executing such approaches, in this way guaranteeing greatest access to employments expected to appreciate fair work.

## 7.7 Standards on Employment Promotion

Standard No. 122 sets out the objective of full, profitable, and openly picked work; other ILO instruments set forward procedures for accomplishing this point. Job administrations (open and private), the work of debilitated persons, little and medium ventures, and cooperatives all have influence in making vocation. ILO measures in these fields give direction on utilizing these methods successfully as a part of request to make employments.

## 7.8 Standards on Wages

People work to earn money. However, in numerous parts of the world, access to satisfactory and consistent wages is not ensured. In various nations, non-instalment of wages has prompted colossal pay overdue debts, and wages are some of the time paid in bonds, made products, or even liquor. Substantial pay unpaid debts have been connected to obligation subjugation and bondage. In different nations, labourers face loss of wages when their manager goes bankrupt. ILO models on wages address these issues by accommodating customary instalment of wages, the altering of the lowest pay permitted by law levels, and the settlement of unpaid wages in the event of boss indebtedness.

## 7.9 Standards on Working Time

The regulation of working time is one of the most seasoned worries of work enactment. As of now in the mid-nineteenth century, it was perceived that working extreme hours represented a risk to specialists' well-being and to their families. The principal ILO Convention, embraced in 1919 (see beneath), constrained hours of work and accommodated sufficient rest periods for specialists. Today, ILO benchmarks on working time give the structure to manage hours of work, day-by-day and week-by-week rest periods, and yearly occasions. These instruments guarantee high profitability while protecting specialists' physical and emotional wellness. Principles on low maintenance work have turned out to be progressively essential instruments for tending to such issues as employment creation and advancing equity among men and ladies.

## 7.10 Standards on Occupational Safety and Health

The ILO Constitution puts forward the rule that specialists ought to be shielded from ailment, infection, and damage emerging from their work. However, for a huge number of labourers, the fact of the matter is altogether different. Consistently, 6300 individuals bite the dust as an after-effect of word-related mishaps or business-related infections—more than 2.3 million passings for every year. 317 million mischances happen at work every year, a large portion of these subsequent in augmented non-appearances from work. The human expense of this day-by-day misfortune is incomprehensible, and the monetary weight of poor word-related security and well-being practices is assessed at 4 for each penny of worldwide GDP every year. Managers confront exorbitant early retirements, loss of gifted staff, non-appearance, and high protection premiums because of business-related mishaps and ailments. However, a hefty portion of these tragedies are preventable through the execution of sound avoidance, reporting, and examination hones. ILO benchmarks on word-related riches and prosperity and they give crucial gadgets to governments, managers, and workers to set up such practices and to oblige most great security at work. In 2003, the ILO received a worldwide system to enhance word-related security and well-being which incorporated the presentation of a preventive security and well-being culture, the advancement and improvement of significant instruments, and specialized help.

## 7.11 Standards on Social Security

A general public that gives security to its residents shields them from war and infection, as well as from the insecurities identified with bringing home the bacon through work. Government disability frameworks accommodate fundamental salary in instances of unemployment, disease and harm, maturity and retirement, invalidity, family obligations, for example, pregnancy and childcare, and loss of the family provider. Such advantages are essential for individual specialists and their families as well as for their groups in general. By giving medicinal services, salary security, and social administrations, government disability improves efficiency and adds to the pride and full acknowledgment of the person.

ILO models on standardized savings accommodate diverse sorts of government disability scope under various monetary frameworks and phases of advancement. Government disability conventions offer an extensive variety of choices and adaptability provisions which permit the objective of general scope to be come to step by step. In a globalizing world, where individuals are progressively presented to worldwide monetary dangers, there is developing awareness of the way that an expansive based national social insurance strategy can give a solid cushion against a number of the negative social impacts of emergencies. Consequently, in 2012, the ILC embraced an imperative new instrument, the Social Protection Floors Recommendation.

## 7.12 Standards on Maternity Protection

Raising a family is a loved objective for some working individuals. However, pregnancy and maternity are a particularly helpless time for working ladies and their families. Eager and nursing moms require unique assurance to avert damage to their or their babies' well-being, and they require satisfactory time to conceive an offspring, to recoup, and to nurture their kids. In the meantime, they additionally oblige insurance to guarantee that they will not lose their employment essentially as a result of pregnancy or maternity clear out. Such security not just guarantees a lady's equivalent access to job; it likewise guarantees the continuation of regularly essential wage which is fundamental for the prosperity of her whole family. Shielding the strength of hopeful and nursing moms and shielding them from occupation segregation are precondition for accomplishing veritable balance of chance and treatment for men and ladies at work and empowering specialists to bring families up in states of security.

# 8    Conclusion

Today, sustainability has turned into a need driven by customer awareness and inclination for sustainable items, consistence standards, and an acknowledgment that to secure the future it is imperative to act today. Life cycle way to deal sustainability in textile and apparel industry involves guaranteeing the three aspects of sustainability—social, economic, and environmental.

Textile and apparel industries are basic parts of the world economy, giving jobs to many millions, generally women labourers in almost two hundred nations. The textile industry is encountering creation and authoritative changes all around, with extending exchange action by adjusting employer–worker relations.

Labour relationship in textile industry is experiencing significant changes. Enhanced plans of credit and aptitude improvement for labourers can possibly expand the commitment of materials in the GDP. The government and the industry need to work together and set up an arrangement of activity that locates key issues and distinguishes and evacuates obstructions to development and sourcing methodologies. Advancements in supply chain administration, proficient administrations, and marking will soon be as a key to sustainable growth in the textile industry. The labourer's welfare plans and new remuneration approaches are to be executed to hold workers. The textile industry is experiencing changes, whereby the administration, labourers, and all associates need be prepared to compete for sustainability.

# References

Kabeer N (2004) Globalization, labour standards, and women's rights: dilemmas of collective (in) action in an interdependent world. Feminist Econ 10(1):3–35

Kakuli A, Risberg V (2012) A lost revolution? empowered but trapped in poverty. Women in the garment industry in Bangladesh want more, Swedwatch Report#47

Malik N (2010) Role of hazard control measure in occupational health and safety in the textile industry of Pakistan. Pak J Agric Sci 47(1):72–76

Padmini DS (2012) Unsafe work environment in garments industries. J Environ Res Dev 7(1A)

Tiwari M (2012) Causes of Musco-Skeletal disorders in textile industry. 1(4):48–50. ISSN 2329-3563

Yunus M, Yamagata T (2012) The garment industry in Bangladesh. Chapter 6. In: Dynamic of the garment industry in low income countries: experiences of Asia and Africa. Chousakenkyu Houkokusho, IDE-JETRO

# Sustainable Practices in Textile Industry: Standards and Certificates

**K. Amutha**

**Abstract** "Sustainability is achieved when all people on Earth can live well without compromising the quality of life for future generations"—Rolf Jucker, "A Vision for a Sustainable University". To sustain is to strengthen or in other words is to support or to make comfortable. Every human has his own choices or preferences and would like to lead a comfortable life. But this comfort should not be achieved at the cost of other's discomfort or sufferings. The concept of sustainable development insists on conservation of resources for the future generations. Nature, earth, biodiversity, and ecosystems need to be sustained for harmonious life at the present as well as the future. Textile Industry is a fast growing industry in the twenty-first century both in terms of production volume and employment, and hence, the industry's impact is huge on the society, economy, and environment. Also, textile industry is a thirsty industry with huge consumer of water, a resource that is becoming scarce day by day. The textile industry is indicted to be one among the most polluting industries, and hence, it becomes mandatory to adopt sustainable practices in order to conserve the Mother Nature. At present, most of the industry practices are modified or transformed in order to achieve sustainability, and for these, standards are indispensable. Sustainable practices in textile industry include using less amounts of water, hazardous chemicals, pesticides, and fertilizers; adopting eco-friendly production processes; using less energy for production processes; and introducing 3 Rs—Reduce, Reuse, and Recycle. Society is also gaining awareness on green consumerism and looking for ecoproducts. Sustainability standards and certificates are concerned with the safety of the consumer, the manufacturer, the society, and the environment at large. The standards are developed after cautious research, and certificates are issued up on proper implementation of these standards. This chapter gives an overview of various standards applicable to the textile industry; certification of textile products; the way the standards lead the industry towards sustainability; the elements of sustainable development, and so on.

K. Amutha (✉)
Department of Textiles and Apparel Design, Bharathiar University, Coimbatore, India
e-mail: ammusuman@rediffmail.com

© Springer Nature Singapore Pte Ltd. 2017
S.S. Muthu (ed.), *Sustainability in the Textile Industry*,
Textile Science and Clothing Technology, DOI 10.1007/978-981-10-2639-3_5

**Keywords** Benchmark · Ecosystem · Nature · Standardization · Sustainable development

# 1 Introduction

This chapter gives a brief idea about the standards and certificates that are applicable to the textile industry. Standardization is the process by which rules and regulations are developed and applied for a particular activity, in a consistent and reliable manner; it is achieved by the cooperation of all stakeholders and for their benefit. Standardization comprises of a chain or cycle of procedures implemented by individuals who work as a group to achieve the committed objective. All these standards guide and lead the industry towards environmental sustainability. Standards help to save cost, time, and energy. They also serve as reliable benchmark by which performance can be compared or judged. International Organization for Standardization (ISO) defines standard as "a document that provides requirements, specifications, guidelines, or characteristics that can be used consistently to ensure that materials, products, processes, and services are fit for their purpose".

Kanti Jasani, president of Performance and Technical Textile Consulting, insists the industry to assume implementation of sustainable practices as yet another quality programme. According to him, sustainable textile products should inevitably be quality products. He also states that quality programmes such as Lean and Six Sigma help the implementers to cut down unnecessary steps in production process and add that these concepts are essential for sustainability.

This chapter focuses on the standards and certificates for sustainable development in the textile industry and across the textile supply chain. The primary motive of these standards is to standardize the procedures and practices followed by the industry, in order to achieve sustainability. The requirements are clearly defined in the standard, and implementation is checked through audits and certified upon compliance.

# 2 Global Organic Textile Standard (GOTS)

GOTS is a standard for processing organic fibres through the entire supply chain, beginning from the harvesting of the raw material (fibre) up to the finished (final) product being sold to the consumer. It is followed by many manufacturers around the world, and hence, the products with GOTS certification are accepted widely in the international market. GOTS certification requires social compliance in addition to the environmental compliance. The dyestuffs, chemicals, and reagents used for dyeing and finishing should meet both environmental and toxicological criteria. It is mandatory for wet processing units to have effluent (wastewater) treatment plant (Fig. 1).

**Fig. 1** GOTS logo. *Source*
www.global-standard.org

## Brief History of GOTS

**May 2005**: First version of GOTS was published,
**Oct 2006**: Commencement of GOTS certification system,
**June 2008**: Second version (2.0) and label of GOTS were introduced,
**March 2011**: Third version (3.0) was published, and
**March 2014**: Fourth version (4.0) was published.

## Product Categories

All the textile and allied materials and products are being classified into the following nineteen categories. This categorization eases the process of certification.

| | | |
|---|---|---|
| 1. Accessories | 8. Ladies' wear | 15. Technical textiles |
| 2. Baby wear | 9. Leisure wear | 16. Toys |
| 3. Children's wear | 10. Men's wear | 17. Underwear |
| 4. Fabrics | 11. Non-wovens | 18. Yarn |
| 5. Garments | 12. Raw fibres | 19. Others |
| 6. Home textiles | 13. Socks | |
| 7. Hygiene products | 14. Sports wear | |

## Fields of Operation

The manufacturing processes are categorized into the following seventeen fields which help in defining the criteria and standards based on the field of operation.

| | | |
|---|---|---|
| 1. Dyeing | 7. Mail order selling | 13. Spinning |
| 2. Exporting | 8. Manufacturing | 14. Storing |
| 3. Finishing | 9. Post-harvest handling | 15. Trading |
| 4. Ginning | 10. Printing | 16. Weaving |
| 5. Importing | 11. Processing (other) | 17. Wet processing |
| 6. Knitting | 12. Retailing | |

## GOTS Labels

GOTS offers two label grades for textile products that are only produced according to the standard (Figs. 2, 3 and 4).

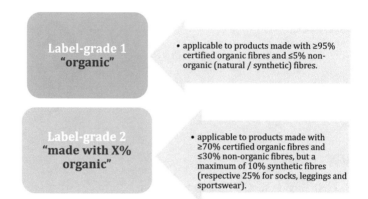

**Fig. 2** Two types of GOTS label

**Fig. 3** GOTS label (organic).
*Source* www.global-standard.
org

The entity who applied for GOTS certification has to undergo inspection by certified bodies, and upon compliance with the standard, the entity would be certified. Then, the final products can be labelled with a suitable label.

**Certification**

Certification process includes on-site inspection and monitoring system that ensures the credibility of the GOTS label. Any textile manufacturing, processing, or trading entity can apply for GOTS certification through any GOTS approved certifying body.

**Fig. 4** GOTS label (x%
organic). *Source* www.global-
standard.org

Using 70 – 94 % organic fibres:

Made with *[x]* %
organic materials

certified by *[certifier's ref.]*
Licence no *[4321]*

**Certifying Bodies**

There are totally eighteen approved certifying bodies for GOTS, spread across the globe. The certifying bodies are independent and special accredited bodies and are authorized to certify on the basis of four different scopes as follows:

1. Mechanical textile processing and manufacturing operations and their products,
2. Wet processing and finishing operations and their products,
3. Trading operations and related products, and
4. Release of positive lists of chemical inputs (dyes and auxiliary agents) to the chemical industry.

**GOTS Monitor (Water/Energy)**

In terms of sustainable development for the textile manufacturers, especially the wet processors, the two most important facets are consumption of water and energy. Hence, it becomes mandatory for the applicant of GOTS certification to provide necessary data about the energy and water resources used for production and their consumption per kilogram of textile output. In order to support and monitor the certified unit, GOTS monitor acts as a valuable tool. The core functions of GOTS Monitor are as follows:

- To offer easy ways to collect data on consumption of energy and water to produce one kilogram of textile material or product.
- To give benchmark values (practical and factory-specific), particularly for wet processing units.
- To offer support to set and monitor rational targets to minimize the consumption of water and energy.

# 3 American National Standards Institute (ANSI) Standards

ANSI was established in the year 1918. It is the official US representative to the ISO. It is a non-profit organization for the US voluntary consensus standards community. It is a neutral forum that ensures reliability of the standards and the compliance evaluation system.

- NSF: National Sanitation Foundation—founded in 1944.
- Renamed as NSF International in 1990 and expanded its services across the globe.
- NSF is a self-governing organization whose mission is to protect and improve global human health.
- It tests the quality and certifies products and establishes standards for food, water, and consumer goods industries.
- NSF is also a third-party certification body for organizations and businesses.
- NSF guides the organizations throughout the certification process to achieve high-quality results in a cost-effective, reliable, and consistent manner.

## 3.1 NSF 140—2015: Sustainability Assessment for Carpet

NSF/ANSI 140 is a leading standard for the carpet industry. This standard is pertinent to evaluate the sustainability and certify the carpet products through their entire life cycle. This standard was developed by the NSF—National Centre for Sustainability Standards (NCSS) through a consensus base public process with a multi-stakeholder group of participants. This standard guides the carpet manufacturing organization towards more sustainable carpet production and defines the performance requirements for both the individual product and the organization as a whole. Prerequisite requirements are being established, and the commercial carpet products are evaluated against these requirements. The evaluation process is based on the life cycle assessment (LCA) principles, and for this, NSF/ANSI 140 makes use of a point system in six key areas (Fig. 5):

**Fig. 5** NSF 140 label.
*Source* www.nsf.org

(i) Public health and environment,
(ii) Energy and energy efficiency,
(iii) Bio-based, recycled content materials, and environmentally preferable materials,
(iv) Product Manufacturing,
(v) Reclamation and end-of-life management, and
(vi) Innovation.

Based on the total point scores, the certificate is issued at three different levels:

- Silver,
- Gold, and
- Platinum.

NSF certified manufacturers are approved to use the NSF Sustainability Certified label on their products as well as for advertisement purpose.

## 3.2 NSF 336—2011: Sustainability Assessment for Commercial Furnishing Fabric

NSF/ANSI 336 is engaged for the evaluation and certification of sustainability of commercial furnishing fabrics through the entire product life cycle. This standard was developed by the NSF–NCSS. The standard was developed by a consensual process with the participation of multi-stakeholder group that includes (Fig. 6):

- manufacturers,
- suppliers,
- regulatory agencies,
- academicians, and
- other industry participants.

NSF/ANSI 336 standard deals with the environmental, economic, and social aspects of furnishing fabric products. This standard is used to assess the product (furnishing fabrics) characteristics in terms of input materials and components, use

**Fig. 6** NSF 336 label.
*Source* www.nsf.org

of water and energy, practices adopted for recycling, and social accountability. NSF/ANSI 336 is based on LCA principles. Accordingly, the furnishing fabrics are evaluated against preset requirements using a point system in eight key areas as mentioned below:

   (i)   Fibre sourcing,
  (ii)   Safety of materials,
 (iii)   Water conservation,
 (iv)   Water quality,
  (v)   Energy,
 (vi)   Air quality,
 (vii)   Recycling practices in manufacturing and end of use, and
(viii)   Social accountability.

This standard is applicable to

- Commercial fabrics generally used in offices, hospitality, healthcare centres, and institutional interiors.
- Woven, non-woven, knitted, bonded, felted, and composite materials used in upholstery, vertical window, furniture system, wall, drapery, cubicle, and top of the bed fabrics.
- Fabrics manufactured in one or multiple amenities and in one or multiple countries (Fig. 7).

Based on the total point scores, the certificate is issued at four different levels:

- Compliance,
- Silver,
- Gold, and
- Platinum.

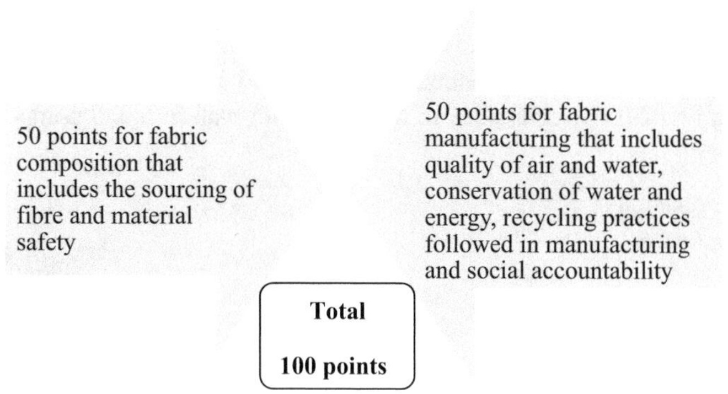

50 points for fabric composition that includes the sourcing of fibre and material safety

50 points for fabric manufacturing that includes quality of air and water, conservation of water and energy, recycling practices followed in manufacturing and social accountability

**Total**

**100 points**

**Fig. 7** Score points for NSF 336 standard

# 4 OEKO-TEX® Certification

OEKO-TEX® is an international association established in 1992 and has 15 textile research and test institutes in Europe and Japan. There are more than 50 local offices spread over 60 countries around the world. The standards are applicable for various textile products to ensure human health. The textile industry is extremely fragmented with each production stage being located at different places of the world, and this complex supply chain calls for some international regulation; hence, Oeko-Tex fulfils it. Oeko-Tex standards are based on the textile ecology which is divided into four segments as follows (Fig. 8).

## 4.1 OEKO-TEX® Standard 100

The OEKO-TEX® Standard 100 is an independent system that tests and certifies the materials or products in the entire textile supply chain that includes raw materials, intermediary, and finished products. All the items are tested for the presence of harmful substances and certified up on compliance with the standard. The criteria catalogue of testing for harmful substances is standardized globally. It is amended and extended on regular basis (Fig. 9).

Materials or products suitable for OEKO-TEX® certification are as follows:

- Greige, dyed, and finished yarns.
- Greige, dyed, and finished fabrics (woven and knit).
- Ready-made articles (all types of clothing, domestic and household textiles, bedlinen, terry cloth items, textile toys, and more).

**Criteria**

Testing for harmful substances includes the following:

- Illegal substances (banned dyes, chemicals, etc.),
- Legally regulated substances,
- Known harmful (but not legally regulated) chemicals, and
- Parameters for health care.

**Fig. 8** Classification of textile ecology

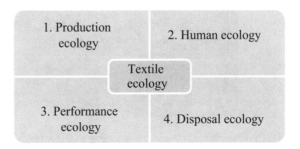

**Fig. 9** OEKO-TEX 100
label. *Source* www.oeko-tex.
com

## Product Classes

OEKO-TEX® Standard 100 employs different test methods for the assessment of harmful substances, and these test methods are chosen based on the end use of the textile. Based on the human ecological requirements, the textile product that comes into close contact with the skin is given more importance. According to this standard, the tested textile products are classified into four different product classes as follows (Fig. 10).

## Certification

The textile product to be certified as per OEKO-TEX® Standard 100 has to be tested for harmful substances, and certification will be done only upon compliance with the required criteria. The testing has to include all the components or materials of the final product which includes fabric, sewing threads, linings, prints, and also non-textile accessories such as buttons, zip fasteners, and rivets. No exception is allowed.

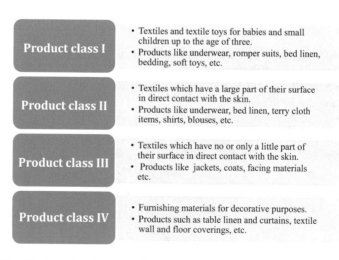

**Fig. 10** Classification of products according to Oeko-Tex 100 Standard

**Modular System**

According to OEKO-TEX® Standard 100, textile products can be tested and certified at any stage of production including accessories manufacturing.

Certificates are issued for:

1. Raw materials,
2. Fibres,
3. Filaments,
4. Yarns,
5. Grey (raw) and finished textile fabrics,
6. Ready-made products (Garments), and
7. Textile and non-textile accessories.

While certifying a finished product, raw materials or intermediate products which have already been tested and certified are exempted from retesting, and for this, the certificate and representative sample materials have to be presented as proof. This modular principle avoids double testing and saves cost and time. It is the responsibility of the manufactures to sustain the human ecology and to maintain the quality of the merchandize by choosing the right ingredients at every stage of production.

**Ready-Made Articles**

Though all the elements have been tested and certified already, consistent with OEKO-TEX® Standard 100, ready-made articles are required to have their own licence number; that is, the final product is marked with neither the several OEKO-TEX® labels nor the certificate number of individual components. The manufacturers of ready-made articles have to assure compliance of their product with the obligation of OEKO-TEX® criteria catalogue in all the individual cases and at the same time integrate these procedures into their normal quality assurance system.

## 4.2 Sustainable Textile Production (STeP)

Sustainable textile production (STeP) has been introduced by OEKO-TEX® certification system for the benefit of manufacturers of branded goods, retailers, and other organizations in the textile supply chain. It serves as a means to communicate the achievements of the textile organizations concerning sustainable production to the public. The certification process is transparent, reliable, and apparent. STeP certification is applicable to all stages of textile goods production that begins from fibre production and extends through the processes of spinning, weaving, knitting, processing (dyeing/printing and finishing), and finally the ready-made textile production (Fig. 11).

**Fig. 11** STeP label. *Source*
www.oeko-tex.com

The main aim of STeP certification is to implement eco-friendly production processes permanently and also ensure optimal health and safety and working conditions that are socially adequate. The STeP standard is vibrant and keeps improving based on the technological changes. Benchmarks are validated frequently, and this helps the certified companies for continual improvement of environmental protection, social responsibility, and efficiency. As a result, the companies are able to attain the best position in the market which is highly competitive.

STeP acts as a source for global brands and retailers to look for suitable suppliers around the world who comply with their requirements concerning environmental protection and social responsibility. This facilitates all the organization in the supply chain to ensure sustainability jointly and also document their commitment clearly and completely.

STeP certification helps to improve the efficiency of the production processes of the textile and clothing manufacturers. It also enables the organizations to find out their position with respect to sustainability and points the areas to be improved. The certification is independent and acts as a proof of sustainable production and builds brand image. It allows the companies to enter into new markets and improves relations through the supply chain.

*Comprehensive approach—textile-specific criteria*

While most of the certification systems account for only individual aspects of sustainability, STeP, in contrast, focuses on widespread, complete analysis and assessment of production processes and conditions in a sustainable manner. Another advantage is that the STeP certification adapts criteria that specially suit the textile industry situations. For wide application of STeP certification, it has to assure that the criteria specified are comparable at international level and standardized accordingly. The standardization process includes continual analysis, evaluation, and updating whenever required, for example, in case of new market development, when legal regulations are passed/changed, upon new scientific or technological innovations.

*Evaluation*

The STeP certification is offered by the institutions who are members of the International OEKO-TEX® Association, and these institutes are spread across the world. There are totally sixteen member institutes and representative offices

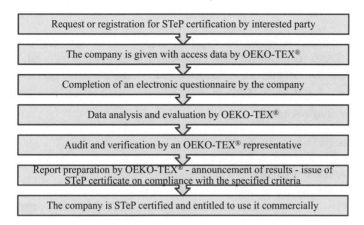

**Fig. 12** STeP certification process

worldwide. These institutes are independent but accredited by OEKO-TEX®
Association to carry out textile testing and certification. They are highly competent
and well experienced in the textile industry.

Certification

The STeP certification process includes the following 7 steps (Fig. 12).

Duration of the Certification

It takes about three months to complete the above seven step process of certification. The duration varies according to the production facility and the individual
situations. For example, if the data could be collected fast, the time taken is less and
vice versa; if certain individual business areas are already certified, then the process
ends faster.

Certification Cost

The expenditure for STeP certification varies based on the following:

- the scope of the company in terms of size and production volume,
- the criteria to be tested at every production stage, and
- prior implementation measures for environmental protection and social
  responsibility by the company.

The certification cost includes expenses for support services rendered by the
OEKO-TEX® institutes during the application, evaluation, audit process (preparation, implementation, and documentation), certificate issue, and also an authorization fee.

Certificate

A STeP certificate is issued for a period of three years and subject to further extension. The prerequisite for STeP certification is that the company has to comply with the minimum requirements specified and the requirements are classified into the following six modules or areas.

### Management of chemicals

- Meet the guiding principle of restricted substances list (RSL),
- Appropriate management for harmful substances,
- Implement the principles of "green chemicals",
- Regular training for handling of the chemicals,
- Commitment to communicate about the risks involved in using chemicals, and
- Supervision of chemicals in use.

### Environmental performance

- Not exceeding the specified limit values,
- Make use of best available production technologies,
- Optimization of production processes,
- Use the resources efficiently,
- Responsible handling of waste, wastewater, emissions, etc., and
- Reduction of the $CO_2$ footprint.

### Environmental management

- Documentation and implementation of a right environmental management system,
- Assurance of environmental targets,
- Conception of environmental reports periodically,
- Appointment of an environmental representative,
- Periodic training for the implementation of environmental targets, and
- Implementation of existing environmental protection systems such as ISO 14001.

### Social responsibility

- Ensuring socially acceptable working conditions as per the UN and ILO conventions,
- Implementation of performance appraisals for all the employees,
- Implementation of existing social standards like SA 8000, and
- Assured training for employees concerning the social issues of an operation.

### Quality management

- Implementation of a suitable quality management system like ISO 9001,
- It is necessary that the flow of goods and products be traceable with proper documentation and responsibility, and
- Use of advanced management practices like risk management or corporate governance.

### Health and safety

- Appropriate measures to ensure health and safety of workers in the workplace (e.g. gloves, masks, filter systems, and ear protection),
- The buildings and production plants are assured for safety (e.g. fire alarms, emergency exits, and restricted areas),
- Measures for prevention of risk, and
- Implementation of existing safety standards like OHSAS 18001.

### Assessment

The companies have to provide a number of data and parameters to OEKO-TEX® for satisfactory evaluation of the sustainability of production processes. Data are being collected through Web, and an evaluation tool is used to assess the data. After applying to an OEKO-TEX® institute, the company receives the necessary access data which enables them to send the information through an online questionnaire. Then, the institute would evaluate the data and send the report in detail about the results. Then, the company can proceed to the next stage by either following the procedure for sustainable production or implementation of the certification process to achieve the STeP label.

### Scoring

STeP certification is offered at three different levels based on the degree of sustainable production and working conditions:

- Level 1 = entry level,
- Level 2 = good implementation with further optimization potential, and
- Level 3 = exemplary implementation.

The assessment results are presented in a detailed manner so that the company can know its position in terms of sustainability. It also demonstrates the particular areas that have additional potential for optimization.

### Audits

The company provides necessary data and information for the assessment, and these data are analysed and evaluated by the OEKO-TEX® institute. OEKO-TEX®

institute has a person in charge to carry out the verification through audit at the production capacity.

The internal quality management of OEKO-TEX® is comprehensive, and it ensures audit results of global standards. For this purpose, the auditors are offered joint training and further education regularly and also the technical OEKO-TEX® expert groups meet annually.

**Verification**

OEKO-TEX® carries out audits at different points of time as follows:

- Initial auditing at the time of application,
- Routine audits at the time of extension,
- Intermediate/compliance audits, and
- Unannounced audits at the production facility.

## 4.3  Made in Green by OEKO-TEX®

**Made in Green** is a label for textiles. It operates independently and is suitable for finished textile goods and semi-finished or intermediate products at all levels of the textile supply chain. It ensures that the products are made of materials free from harmful substances and the processes employed for manufacturing are environmental friendly, and the working conditions are safe and socially responsible. This single label replaces both the OEKO-TEX® Standard 100 and Made in Green by Aitex. This label is applicable to any kind of clothing and furnishings; either the item is a finished or semi-finished product at any level of the textile supply chain (Fig. 13).

The Made in Green label by OEKO-TEX® confirms that the product carrying this label is tested for harmful substances and found to be within the tolerable limits; also, sustainable production processes are followed as per the guidelines of OEKO-TEX®. Every piece of article carrying the Made in Green label possesses a product ID which is unique and also serves as a means of traceability and transparency to the consumer. It gives details such as the production facility employed, the stage of production, and the countries where the manufacturing processes were carried out.

**Fig. 13**  Made in Green label.
*Source* www.oeko-tex.com

**Validity**

The Made in Green label is valid for a year after which it must be reissued.

**Additional Benefits**

- This label could be used directly by innovative leaders for the promotion of new functions and features of their products and as a means to emphasize their close partnership with suppliers.
- A key factor for the successful implementation of sustainability is the reliable relationship between the supplier and the manufacturer. The sustainability criteria include quality of the product, educating and training the employees at regular intervals, and improved working environment. The Made in Green label conveys all these to the customer.
- Made in Green label makes the entire supply chain of the company transparent and acts as a sign of trust that the suppliers from other countries also stick to the quality standards as per the guidelines and work principles.
- Though the Made in Green label is a mark of the existing sustainability of the supply chain, the companies may work further towards continuous improvement for achieving greater sustainability at individual production units.

**Control Tests**

In cases where all the textiles are certified by OEKO-TEX® Standard 100, control tests are performed before the products are labelled with Made in Green label. In addition to this, random checking is done during unannounced company audits. These unannounced company audits are performed as a measure to verify whether the existing STeP certification complies with both the environmental and social criteria.

## 5   ISO 14000 Environmental Management

The ISO 14000 is a collection of standards that offers practical tools for manufacturers or businesses of all categories who are interested to improve their environmental management responsibilities. The chief standard ISO 14001:2015 and its supporting standards like ISO 14006:2011 are developed in order to achieve this. There are many other standards in the family whose focus is on specific activities such as audits, communications, labelling and life cycle analysis, and also environmental challenges like climate change (Fig. 14).

**Fig. 14** ISO 14001 label.
*Source* www.iso.org

## 5.1 ISO 14001 Environmental Management System

ISO 14001 is a global standard for implementation of environmental management systems and is based on the four-step method called Plan–Do–Check–Act (PDCA cycle or Deming cycle).

### ISO 14001 Certification

Any organization interested to be certified by ISO 14001: environmental management system (EMS) should comply with all the clauses of the standard. For this, the company has to develop guidelines, set of rules, and procedures in order to make sure that it has only minimum or less impact on the environment.

### Benefits of ISO 14001 Certification

ISO 14001 certification of any organization assures that it meets the international environmental standards as stipulated by environmental management system (EMS) which is industry-specific in nature. This standard and certification are applicable to any company irrespective of the size, large or small; both manufacturing and service sectors; traders; and different types of industries.

### Key Benefits of ISO 14001

- Help in management of resources, energy, and waste as a measure of cost reduction,
- Reliable and build a corporate image,
- The impact on the environment is measured, scrutinized, and controlled,
- Comply with the regulations and create awareness,
- The environmental performance of the whole supply chain improves,
- Serve as a protective tool,
- Cut down the insurance costs, and
- Attract customers and business partners.

### ISO 14001 Certification

The ISO 14001 EMS certification is accomplished through an audit process which comprises of two stages:

- ISO 14001 Stage 1: Preassessment
- ISO 14001 Stage 2: Certification

ISO 14001 EMS certificate will be issued only on successful completion of the audit process. The certificate once issued is valid for three years, and every year mandatory audits are conducted in order to ensure compliance. By the end of the valid period, a reassessment audit is required to get certified for the next span of three years.

The ISO 14001 EMS certification requires a company to have the following ISO 14001 compliant documents, procedures, processes, and policies:

- Environmental manual,
- Environmental policy,
- Environmental aspect and impact procedure,
- Environmental legal and other requirements procedure,
- Environmental objectives, targets, and programmes,
- Adequate resources to maintain the EMS,
- Communication, documentation, and record procedures,
- Environmental compliance and internal audit procedures,
- ISO 14001 operational control and emergency procedures,
- Environmental monitoring,
- ISO 14001 nonconformity, corrective, and preventative action, and
- EMS management review.

## 5.2  ISO 14020 Environmental Labelling

ISO 14020 consists of a series of standards that provides global recognition for business organizations. They also offer credible international benchmarks against which the organizations set standards for labelling of their products (Fig. 15).

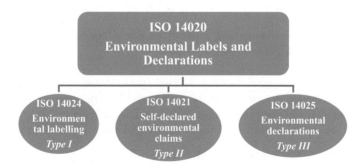

**Fig. 15**  ISO 14020 group of standards

### 5.2.1   ISO 14024 Environmental Labels and Declarations: Environmental Labelling—Principles and Procedures

ISO 14024 is a global standard established for environmental labelling which includes selection of:

- Product categories,
- Product environmental criteria, and
- Product function characteristics.

It is commonly termed as "ecolabelling". This standard serves as a benchmark for the Global Ecolabelling Network (GEN). The selection of product categories for ecolabelling begins with a feasibility study for establishing the product categories and then the preparation of proposal (specifically for the interested parties) which summarizes all its components, findings, and the considerations. The selection of product environmental criteria deals with the decisions on final criteria based on the standard which links the stages of product life cycle and the major environmental indicators. During this process, factors such as neighbouring, provincial and international environmental issues, existing technology, and economic aspects are taken into consideration. The process of ecolabelling comprises of:

(i)   Identification of the product life cycle stages based on the environmental impacts and analysis to ensure adequacy.

(ii)  Allotment of weighting factors for the selected environmental requirements and statement of explanation and justification for each weighting factor.

(iii) Determination of the criteria that accurately reflect the particular environmental aspects and assignment of numerical values to them using either scale point system or other relevant approaches.

(iv)  Stipulation of test methods for any given criteria and examination of the available laboratories capable of performing the tests.

### 5.2.2   ISO 14021 Environmental Labels and Declarations: Self-Declared Environmental Claims

ISO 14021 finds wide applications that deal with the voluntary, self-declarative, and environmental claims for not only manufactured products but also services such as banking and tourism. The chief goal of ISO 14021 is the harmonization of the use of environmental claims (self-declared) with the following expected benefits:

- Precise and provable environmental claims,
- Encouragement for improvement of the environment,
- Avoiding unjustifiable claims,
- Lessening uncertainty in market,
- Assistance in international trade, and
- Variety of choice for consumers.

The basic rules for environmental claims are expressed through three key elements as follows:

1. *Use of symbols*: the claim is not just expressed as text but also with pictures, symbols, or logos.
2. *Evaluation and claim verification requirements*: it is indispensable to verify the claims before they are made and is made accessible to anyone on request.
3. *Specific requirements for selected claims*: it is suitable for certain claims that are used frequently, e.g. recyclable and biodegradable.

**Specific Requirements for Selected Claims**

The final element of the standard is the specific requirements for selected claims, for example:

- Degradable,
- Recyclable,
- Recycled content, and
- Reduced energy/water consumption.

The standard was originally published in 1999 and was amended/revised and published in 2011 that reflects the marketplace developments with the addition of the following terms, qualifications, and evaluation methodology:

- Renewable material,
- Renewable energy,
- Sustainable,
- Claims relating to greenhouse gas emissions,
- Product "Carbon footprint", and
- "Carbon neutral".

### 5.2.3 ISO 14025 Environmental Labels and Declarations: Environmental Declarations—Principles and Procedures

ISO 14025 is a standard of Type III environmental declaration which takes life cycle data as the basis for the establishment of principles and procedures. A Type III environmental declaration can be described as computation of environmental data for a product with preset categories of parameters based on the ISO 14040 series of standards, but not excluding additional environmental information. Type III environmental declarations state the environmental performance of a product which facilitates comparison with other products of similar functionality. These declarations:

- Are based on verified LCA data, life cycle inventory analysis (LCI) data, and life cycle impact assessment (LCIA) of a product.
- Are developed using predetermined parameters.
- Are subject to the administration of a programme operator.

## 5.3  *ISO 14044 Environmental Management—Life Cycle Assessment*

ISO 14044:2006 is an international standard that specifies requirements and provides guidelines for life cycle assessment (LCA) including the following:

- definition of the goal and scope of the LCA,
- the life cycle inventory analysis (LCI) phase,
- the life cycle impact assessment (LCIA) phase,
- the life cycle interpretation phase,
- reporting and critical review of the LCA,
- limitations of the LCA,
- relationship between the LCA phases, and
- conditions for use of value choices and optional elements.

This standard is applicable to both manufactured products (termed as goods), including the raw materials and the intermediate products and services. The three key criteria for LCA are mass, energy, and environmental significance.

## 6  Textile Exchange (TE) Standards

Textile Exchange is an international, non-profit organization whose mission is to motivate and organize people to speed up sustainable practices in the textile value chain and transform the textile industry towards sustainability. The TE focuses on minimization of harmful impacts of the textile industry across the globe and maximization of the positive effects. It works on identifying the best practices for the textile chain in order to minimize the impacts on human and environment. Companies willing to be more sustainable may join the TE.

## 6.1  *CCS—Content Claim Standard*

This is a voluntary, international standard which acts as a tool to validate content claims of textile products. The core objective of the Content Claim Standard (CCS) is to confirm or verify the accuracy of content claims (Fig. 16).

**Fig. 16** CCS label. *Source*
www.textileexchange.org

1. CCS ascertains the nature and quantity of the raw materials used for production so that the final product content is claimed accordingly.
2. CCS is applicable to all the products whose content varies from 5 to 100 % of the claim.
3. CCS is versatile and suitable for any material along any supply chain.
4. CCS protect the reliability and uniqueness of the claimed material.
5. The standard gives chain of custody certification.

The Content Claim Standard is applicable to organizations who would like to exhibit the legitimacy and validity of the content of the material used to make products. The organization may be the manufacturer, buyer, or seller of the product. The standard offers verification by third-party organization for the content claim of the product. It also pays attention to the flow of materials or goods inside the production facility and also goods purchased from other sources. The standard includes processes such as manufacturing, storage, handling, and shipping. The CCS should be used in cases where a claim cannot be backed up by another standard or recognized testing method, or when other verification methods are not in place (internal track and track systems, genetic markers, etc.).

## 6.2 OCS—Organic Content Standard

Textile Exchange is the publisher of the Organic Content Claim (OCS). This standard acts as a replacement for the OE 100 and OE blended standards from 19 may 2013. ECOCERT purposes the certification of textiles made with organic grown materials according to the Organic Content Standard. The objective of this standard is to assure the traceability and reliability of the raw materials during all the manufacturing stages (Figs. 17 and 18).

**Fig. 17** OCS label. *Source*
www.textileexchange.org

**Fig. 18** OCS (blended) label.
*Source* www.textileexchange.
org

Two different labels are offered by the standard according to the composition.

1. OCS 100 logo is used for only product that contains 95 % or more organic material.
2. OCS blended is used for products that contain 5 % minimum of organic material blended with conventional or synthetic raw materials.

## 6.3 RDS—Responsible Down Standard

The textile industry has become more and more aware of the challenges in sourcing. Particularly, sourcing is complex for animal products such as feather and down, wool, angora, cashmere, and leather. To make the sourcing easier and more sustainable, the Responsible Down Standard (RDS) was developed and it serves to be a primary standard for animal welfare in down and feather products. With the global expertise of Control Union, services are offered for all animal products integrated in textiles (Fig. 19).

The standard is owned and managed by Textile Exchange. The RDS has been developed by Control Union, The North Face, and Textile Exchange through a wide stakeholder review process, which involved among others worldwide down supply chain members and major animal welfare organizations. The standard development process included pilot audits at all levels of a typical down supply chain and therefore resulted in a scalable standard enabling worldwide applicability by any type of operation (farm, slaughterhouse, processor, trader, collector, garment manufacturer, etc.) in the down supply chain.

**Fig. 19** RDS label. *Source* www.textileexchange.org

The scope of the RDS includes the complete down supply chain—from the farms and slaughter facilities (animal welfare) to the down processors and garment factories (traceability).

Down certified according to RDS shall:

Not be sourced from waterfowl which were force-fed and/or live-plucked,
Be sourced from waterfowl raised respecting animal welfare aspects based on the five freedoms for farm animals, and
Include transparency along the complete supply chain with an integrated traceability.

## 6.4  GRS—Global Recycle Standard

The Global Recycle Standard (GRS) was originated by Control Union and now managed by Textile Exchange. This standard is proposed to establish independently verified claims as to the amount of recycled content in a yarn. Additional importance is given in prohibition of specific chemicals, materials that need water treatment, and safeguarding the rights of workers and also inclusion of weaver into the standards (Fig. 20).

- It is mandatory to keep full records of the use of chemicals, energy, water consumption, and wastewater treatment including the disposal of sludge;
- All prohibited chemicals listed in GOTS are also prohibited in the GRS;
- All wastewater must be treated for pH, temperature, COD, and BOD before disposal; and
- More importance being given to worker's rights.

The GRS provides a track and trace certification system to ensure that the claims made about a product can be officially backed up. It consists of a three-tiered system:

- Gold standard—products contain between 95 and 100 % recycled material;
- Silver standard—products contain between 70 and 95 % recycled product;
- Bronze standard—products have a minimum of 30 % recycled content.

**Fig. 20** GRS label. *Source* www.textileexchange.org

## 6.5  RCS—Recycled Content Standard

The RCS is used as a chain of custody standard to keep track of recycled raw materials through the supply chain. The standard was developed by the Materials Traceability Working Group, part of OIA's Sustainability Working Group. The RCS uses the chain of custody requirements of the Content Claim Standard (Fig. 21).

The RCS uses the ISO 14021 definition of recycled content, with interpretations based on the US Federal Trade Commission Green Guides, the intention of which is to comply with the most widely recognized and stringent definitions. Sellers of RCS products are advised to reference the allowed recycled content claims in the countries of sale, to ensure that they are meeting all legal product claim requirements.

The RCS does not address other inputs, environmental aspects of processing (such as energy, water, or chemical use), any quality or social issues, or legal compliance. Intended users of the RCS are recyclers, manufacturers, brands and retailers, certification bodies, and organizations supporting recycled material initiatives.

## 6.6  RWS—Responsible Wool Standard

Wool is a significant fibre in the textile industry with a long history and an even longer future. It is a versatile fibre with good comfort and performance character-istics. Hence, it has a wide range of applications and preferred by most consumers. Wool owes its unique properties to the sheep that grow it, and we owe it to the sheep to ensure that their welfare is being protected. To this end, Textile Exchange is developing the Responsible Wool Standard.

The RWS is being developed through an open, multi-stakeholder process. The International Working Group represents the broad spectrum of interested parties, including animal welfare groups, brands, farmers, wool suppliers, and supply industry associations, covering both apparel and home categories. The goals of the Responsible Wool Standard are to provide the industry with the best possible tool to:

**Fig. 21** RCS label. *Source* www.textileexchange.org

- Recognize the best practices of farmers around the globe,
- Create an industry benchmark for animal care and land management to drive improvement where needed,
- Ensure that wool comes from farms with a progressive approach to managing their land and from sheep that have been treated responsibly,
- Provide a robust chain of custody system from farm to final product so that consumers are confident that the wool in the products they choose is truly RWS.

# 7  Cradle to Cradle Certified Standard

Cradle to Cradle Certified$^{TM}$ is a standard for assessment of product quality in terms of product design and safety for the consumer and the environment, by multiple aspect approach. The evaluation process includes both the materials used and the manufacturing practices on the basis of the following five attributes (Fig. 22):

1. Material health,
2. Material reutilization,
3. Renewable energy and carbon management,
4. Water stewardship, and
5. Social fairness.

The core objectives of Cradle to Cradle® design are waste elimination and product development for closed-loop system. The certification validates by external means, and the certificate is issued at five different levels:

- Basic (provisional),
- Bronze,
- Silver,
- Gold, and
- Platinum.

**Fig. 22** Cradle to Cradle label. *Source* www. c2ccertified.org

There are totally 10 product categories under Cradle to Cradle certification and 21 products are being certified under the category "Fashion + Textiles". The products include fibre, yarn, fabric, thread and trims, dyes, and apparel.

## 8 Advantages and Disadvantages of Sustainability Standards

Many national and international organizations are involved in the development of standards, and mostly the certification is done by third-party organizations who are accredited by the organization that develops standards. The sustainability standards discussed above have both advantages and disadvantages.

Advantages are as follows:

1. Standards aid in minimizing the waste.
2. Standards lead towards sustainable development.
3. Standards help to increase productivity.
4. Standards ensure safety.
5. Standards facilitate in cutting down costs.
6. In the era of fast changing technology, standards make innovations easier.

Disadvantages are as follows:

1. There are many standards, and it may be complicated for the organization to choose a suitable standard.
2. Standards compel the organization to change their methods.
3. The implementation of standards results in some redundant actions that lead to loss in productivity.
4. It takes extra amounts of time, money, and paperwork for documentation and implementation of the standard.

## 9 Conclusion

Sustainability standards help industries to eliminate greenwashing, lower the investment risks in green innovations, and speed up the evolution to a sustainable future. They also help industries to save money by means of adopting more sustainable operational practices and business approaches. They facilitate industries to compete and sell their products in the international market. Any product with sustainability logo or mark has a special concern in the market. Since there are different ways or methods available for production, the ideal method may be chosen based on standards. Standards help the manufacturers to be in pace with the growing competition at the global level. Standards ensure that products are of good quality and mitigate the costs of trial and error, and research and development. Above all, standards aid to minimize the negative impacts posed on the environment. Standards are essential for the globalized era.

# References

Schmitz-Hoffmann C, Schmidt M, Hansmann B, Palekhov D (2014) Voluntary standard systems: a contribution to sustainable development. Springer, Berlin

Muthu SS (2014) Roadmap to sustainable textiles and clothing: regulatory aspects and sustainability standards of textiles and the clothing supply chain. Springer, Berlin

Wilbanks A, Oxford N, Miller D, Coleman S (2009) Textiles for residential and commercial interiors, 3rd edn. London: A&C Black

http://ec.europa.eu/environment/industry/retail/pdf/issue_paper_textiles.pdf

https://www.forumforthefuture.org/sites/default/files/project/downloads/fashionsustain.pdf

https://unfss.files.wordpress.com/2013/02/unfss_vss-flagshipreportpart1-issues-draft1.pdf

www.iso.org

www.oeko-tex.com

www.aatcc.org

www.textileexchange.org

www.global-standard.org

www.nsf.org

www.madeingreen.com

www.responsiblewool.org

www.c2ccertified.org

# Sustainable Design and Business Models in Textile and Fashion Industry

**Rudrajeet Pal**

**Abstract** Textile, clothing, and fashion (TCF) are one of the most unsustainable industries in the world. This challenges triple-bottom-line sustainability, thus calling for increased intervention by designing sustainable development. Several industrial sustainability models have addressed this issue, but they assume incremental improvements and growth while addressing global challenges. Thus, a sustainable business model perspective is required to think and go beyond these increments and reconceive radically how businesses should operate to drive system-level sustainability. In-line with 8 major sustainable business model archetypes existing, this chapter first contextualizes them in TCF industries and goes further to identify 5 key design elements (and underlying strategies) underpinning them. The knowledge of these key design elements (product, process, value network, relation, and consumption pattern), upholding a system thinking approach, will further assist both research and practice to strategically develop and improve the sustainable innovations and business models.

**Keywords** Sustainable design · Sustainable business models · Textile and clothing · Fashion

## 1 Introduction

Textile, clothing, and fashion (TCF) are one of the most polluting and resource-draining industries in the world, next only to oil, in terms of environmental impact (DEFRA 2008; Fletcher 2008). This is driven largely by environmentally unfriendly production practices, involving large-scale destruction of the ecosystem, viz. widespread use of chemicals in the production and processing of textiles and clothing, emission of harmful greenhouse gases, and other effluents followed by

R. Pal (✉)
Department of Business Administration and Textile Management,
Swedish School of Textiles, University of Borås, Borås, Sweden
e-mail: rudrajeet.pal@hb.se

© Springer Nature Singapore Pte Ltd. 2017
S.S. Muthu (ed.), *Sustainability in the Textile Industry*,
Textile Science and Clothing Technology, DOI 10.1007/978-981-10-2639-3_6

throwing of thousands of tons of textiles every year for either landfill or inciner-ation. In addition, in order to meet the competitive pressures of current TCF industry, manufacturing is largely concentrated in low cost bases where there are neither stringent pollution controls nor regulations (Hethorn and Ulasewicz 2008), along with lack of compliance with social aspects such as fair labor and health and safety standards. Moreover, the TCF industries are one of the worst in terms of promoting materialism, increasing driven by a system of classical market economy, providing the basis for emergence of a throwaway society that is based on economies of scale, planned obsolescence of products, short product life cycle, and consequently ever-growing demand of consumers for new products and services (Mont 2008). This to a large extent has resulted in challenges to triple-bottom-line (TBL) sustainability by promoting "producers continue producing and consumers continue buying" (Fuad-Luke 2009), thus calling for increased intervention by designing sustainable development by understanding:

1. Increased effects in terms of TBL: economic, resource constraints, pollution, and social impacts.
2. Need for high degrees of material and resource efficiency.
3. Need for shifting from linear to circular or closed loop—why a focus toward circular economy (CE)

## 2 Sustainable Development

The demands for creating greater environmental and social value, while delivering economic sustainability, suggest the key to radically improve sustainable perfor-mance through sustainable development. Sustainable development is "development that meets the needs of the present without compromising the ability of the future generations to meet their own needs" (Brundtland Report 1987), thus implying needs to attain TBL sustainability. Sustainable development in TCF industries is an approach intended to minimize the "negative environmental impacts" of the current fashion system and, in turn, maximize the positive impacts (benefits) for the society all along the value chain, thus creating sustainable value (Gardetti 2016). To achieve sustainable development, several industrial sustainability models and frameworks are present, viz. industrial ecology, cradle-to-cradle, and The Natural Step (Evans et al. 2009). Aneja and Pal (2015) have further highlighted and compared eight major sustainable development frameworks and their strategic vectors in context to the textile industry. However, Bocken et al. (2013, 2014) points out the inherent shortcomings of these "industrial sustainability" models, as they are broadly underpinned by assumptions of incremental improvement and growth while addressing global challenges. Bocken et al. (2014) further state "to deliver long-term sustainability requires fundamental changes in the global indus-trial system, and this necessitates an integrated approach that goes beyond just ecoefficiency initiatives and reconceive how businesses operate."

For this, sustainable enterprises and their representative business models provide the potential for a new approach and can serve as a vehicle to drive societal innovation (e.g., poverty reduction and well-being), and along with preserve ecological integrity driven by technological innovation (Hart and Milstein 2003), thus leading to a system-level sustainability. Thus, sustainable business models (SBMs) provide a radical system-level perspective toward creating new systems.

In this context, this chapter will discuss the SBMs that uphold sustainability in the TCF industries, in accordance with the 8 key archetypes proposed in Bocken et al.'s (2014) (see Sect. 3). These archetypes are deemed integral in industrial systems for identifying and understanding the key design aspects or elements underpinning these SBMs, and how they are instrumental in leading toward sustainable development in TCF industries.

Hence, this chapter will include the following:

- Major sustainability archetypes and business models (in-line with Bocken et al. (2014) SBM archetype framework), in TCF industrial value chains,
- Key research and company-driven initiatives taken along these archetypes, and
- Sustainable design aspects/elements in these archetypes from a system thinking perspective.

# 3 Sustainable Business Models and Archetypes

Sustainable business models (SBMs) are defined by Bocken et al. (2014) as those that create significant positive and/or significantly reduced negative impacts for the environment and/or society, through changes in the way the organization and its value network create, deliver, and capture value (i.e., create economic value) or change their value propositions. Thus, they go beyond delivering just economic value to include solutions that generate environmental and social values as well (build on a TBL approach) for a broader range of stakeholders using both systems and firm-level perspectives (Lüdeke-Freund 2010; Stubbs and Cocklin 2008).

Bocken et al. (2014) have identified eight different categories of SBM archetypes, describing the underlying mechanisms and solutions that contribute toward designing transformational innovations. These SBMs include closed-loop business models, natural capitalism, social enterprises, product-service systems (PSSs), and other new economic concepts (Bocken et al. 2014) and are as follows:

1. Maximize material and energy efficiency,
2. Create value from "waste,"
3. Substitute with renewables and natural processes,
4. Deliver functionality rather than ownership,
5. Adopt a stewardship role,

6. Encourage sufficiency,
7. Repurpose the business for society/environment, and
8. Develop scale-up solutions.

The criteria for selecting and categorizing various examples to construct the eight archetypes are based upon a methodological framework by highlighting the similarities and differences and the constant comparison of these exemplar types, along with the constant comparison of low level codes (e.g., specific company innovations and initiatives) and higher-level codes as prescribed in Boons and Lüdeke-Freund (2013), viz. social, technological, and organizational innovations [see Bocken et al. (2014) for details].

The next section illustrates these archetypes, their key characteristics and design elements by using exemplar cases (evident in TCF industries), thus underpinning the sustainable practices of each SBM archetype. These cases must not be considered exhaustive but only representative of the major types of SBMs under each archetype and their design elements (Fig. 1). Real-world examples are used to depict these archetypes and SBMs.

## 3.1 Archetype 1—Maximize Material and Energy Efficiency

This archetype, in comparison with higher resource throughput and consumption, aims for maximizing material productivity and energy efficiency to lower the resource consumption by reducing the volume of resource flow (Stahel 2007). Implementation of such strategies encompasses concepts such as designing of ecoproducts, lean and clean production approaches, and waste reduction.

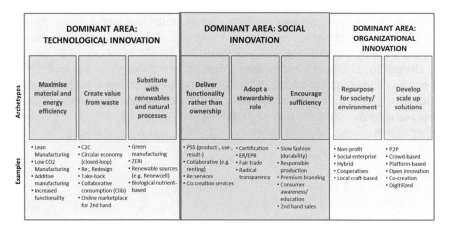

**Fig. 1** SBM archetypes and business cases [adapted from Bocken et al. (2014)]

Considering the increasing volatility in the energy prices, ecodesign aims at improving the energy efficiency and resource effectiveness (closing the material loop) by implementing various strategies, viz. dematerialization or multi-functionality. From the marketing perspective, the current Ecodesign Directive set by European Union (EU) can be extended to cover design, in terms of both scope and markets covered (from energy-related products to all products and services). These products are focused toward limited usage of energy, reduced $CO_2$ emissions, minimized negative environmental impact through energy focus (De Groene Zaak and Ethica 2015), considering the fact that the product design and development plays a significant role in determining the environmental impact of textiles (accounting for nearly 80 % of the total) during various lifecycle stages (Norden 2015). Such ecodesign principles include not only the choice of the material but also the functionality of the product throughout the life cycle (effecting the environment in terms of water and energy consumption). For instance, Continental Clothing—a UK-based brand—has developed an EarthPositive Apparel that is 100 % organic with 90 % reduced $CO_2$. This reduced impact (an EarthPositive T-shirt saves around 7 kg of $CO_2$) is achieved through combination of low-impact organic farming, efficiency in manufacturing and transportation, and the use of renewable energy instead of the fossil fuel-based grid electricity.

In practice, many forerunner companies have started working beyond ecodesign by embracing circular design principles, aimed to go beyond reducing the negative environmental impacts and succeed in creating a positive regenerative impact (De Groene Zaak and Ethica 2015). Gwilt (2014) illustrates using examples various concepts, such as mono-materiality, modular design, that have the potential to design for circularity. Use of mono-materials in products has been considered strategic due to the higher scope of recycling, which is lost in case of blends and also due to certain finishing steps adding contamination into the products. Gwilt (2014) highlights the need to experiment with different embellishment techniques, such as laser-cutting or needle-punching, which can provide detail without contaminating the fiber. This further creates degrees of modularity in the product which is beneficial for easy separation and recycling. Additive manufacturing in this context is a disruptive technology that can be explored in apparel manufacturing; recently, 3D laser printing technology has been used by fashion designer Iris van Harpen and architect Julia Koemer, in collaboration with a Belgium company (Materialise) to fuse small particles of plastic and print continuous surfaces without seams into laces. N12 is a ready-to-wear bikini collection made from 3D printed Nylon 12. Such innovations could be a game-changer in the global sports footwear market as well.

Modular garment design is yet another solution to develop a range of detachable features that can facilitate replacement or repair and in turn can reduce the effect on the environment in terms of water and energy consumption, as well as $CO_2$ emissions in the use phase. These conditional product design strategies are necessary to explore in terms of their environmental impact, as a logical route leading to circularity. At the usage stage, adding functionality reduces the total number of products in use and also enhances the active lifetime of each product. There are different ways to increase this active use time of a product and largely falls under the Archetype 6: "encouraging sufficiency," discussed later.

Further from the process design perspective, lean and green manufacturing seeks to improve resource efficiency and reduce waste and emissions through product-process design.

## 3.2 Archetype 2—Create Value from "Waste"

In contrast to today's linear "take, make, waste" economic model, the concept of creating value from "waste" is underpinned by the idea of restoration rather than disposability, by designing and optimizing products, components, and materials for multiple cycles of disassembly and reuse.

Such closed material and energy loops imply that the materials are reused again, either as bulk material, or products, or as components through specific processes (or economic activities), such as refurbishment or recycling, thus influencing the essence of a circular economy (CE) through new ways of production, distribution, and consumption of goods and services (Ellen McArthur Foundation 2013).

Most important implications from "closing the loop" activities are the potential minimization of extraction of materials from nature and the reduction of waste emission to nature. Problems with extraction, such as resource scarcity, and with emissions, such as environmental impact, can potentially be solved. CE also builds on ideas of generating better economic performance with new (circular) business models along these closed loops that focus on selling services instead of products to product life extension (via remanufacture, resell, repair) to lower resource usage (Wijkman and Rockstrom 2012), thus having the potential to generate an economic growth between 1 and 4 % in many Western economies in the near future (ING Economics Department 2015).

Bocken et al. (2014) have highlighted several exemplar SBM types underpinning the central notion of this archetype, viz. circular economy, cradle-to-cradle, reuse, recycle, remanufacture, sharing, and collaborative consumption. It can be noted that all these exemplary types aim toward attaining various degrees of circularity, through various "closing the loop" activities, thus underpinning the concept of CE. The concept of CE goes beyond just recycling and encompass a holistic view toward "closing the loop" along the five major underlying business models: (i) circular supplies, (ii) resource recovery, (iii) product life extension, (iv) sharing platforms, and (v) product as service (ING Economics Department 2015). The inner loops (sharing platforms and product as service) provide the possibility to retain higher value of the original product predominantly centered on service design. Archetype 4 (Deliver functionality rather than ownership) provides detail of various approaches underlying these loops, e.g., shared ownership, peer-to-peer sharing, and collaborative consumption. However, such business models do create the opportunity to design products such that they ensure longevity, durability, ease of maintenance and repair, and if required upgradability (De Groene Zaak and Ethica 2015). This further demands long-term relationship with clients/users (through access to service) and suppliers (by ensuring safe and healthy working conditions and fair wage), as can be seen in case of Mud Jeans, a Dutch Fashion House with strong focus toward "product as service" via leasing of its jeans to clients/users.

On the other hand, along the outer loops of product life extension, value could be maintained or created by repairing, upgrading, remanufacturing, or remarketing of products. In these loops, value is predominantly restored or added by sufficient rework on the products; hence, the original value retention of the product is low, thus demanding higher labor intensity to recreate "new" value (Stahel 2007).

Various reuse- and resell-based business models can be identified in the secondhand clothing sector. Pal (2015) has identified 8 different resell-based business model types in used clothing networks in Sweden, these are:

(i)  Collection-based, like fashion retailers (with or without supply chain partners) engaged with take-back and "shwopping" schemes through their retailer shops, e.g., H&M in collaboration with I: Collect.

(ii)  Direct reselling, when the retailers collect only its own brands and partly resell them through own shops, e.g., Swedish fashion brands such as Boomerang, Filippa K.

(iii)  Business-to-business (B2B) reselling, when the retailers collect but sells to other actors in the chain, e.g., to secondhand retailers.

(iv)  Charities, those have a hybrid business model and partly resells the collected clothes, e.g., Red Cross. In the Nordic countries, the charities are the largest collector of secondhand clothes and nearly 20 % of it is resold by them through their own shops while nearly 50–70 % is exported to various destinations (Eastern Europe, Asia, and Africa depending upon recovered quality).

(v)  Secondhand retailers, function more or less the same way as the charities, except for the fact that they are totally commercially oriented. Similar to charities, they also engage in partnerships with various actors in the network, such as with fashion retailers, charities, and refurbishers, e.g., laundries.

(vi)  Redesign brands, which have the potential to offer higher value-added used clothing through high degrees of redesign and reconstruction of the old clothes.

(vii)  Reclaimers, which mainly collect and resell leftovers from fashion retailers, or sometimes fashion retailers do it themselves by selling through factory outlets, e.g., Branting is a Swedish brand which debrands its leftovers and sales them.

In most cases, reuse and resell business models account for reducing carbon footprint of new garment manufacturing and energy usage by displacing the production of "new." Swedish Environmental Protection Agency (2016) has identified the effects of such displacement, in terms of reduction in carbon footprints by about 1.5, 1, and 0.5 person equivalents/ton, and reduction in primary energy usage by 2.5, 1.75, and 1 person equivalents/ton, for substitution by factors of 1, 0.66, and 0.33 respectively.

The highest possibility for value creation can be achieved via remanufactured fashion, which aims at remaking used clothes through various redesign possibilities so that it at least equals to newly manufactured garments in terms of quality (Sinha et al. 2015). Many such initiatives have started worldwide and are predominantly led by niche and small-scale redesign brands. However, what differentiates remanufactured fashion from that of upcycling is the focus toward process industrialization compared to that being craft-based in case of upcycling design (Sinha et al. 2015). However, the desirability of remanufacturing fashion is high considering the degree of value addition, scope to create employment, and lower use of energy and material. Recent initiatives, e.g., Retextile (2016), in Sweden are working along this direction to develop redesign-make strategies and methodologies.

Another key aspect for initiating these circular loops based on product life extension pivots onto the reverse logistics operations of collection and disposition of the used products. Sorting or disposition is a key issue and also barrier to circularity of textiles and clothing, particularly due to complex blends of fibers and various degrees of contamination throughout the product life cycles. To ensure better recyclability, various ongoing initiatives aim at developing new mechanical and chemical sorting techniques. Mechanical sorting techniques include use of barcodes (for checking productivity of manual sorting, or price tagging), RFID (in retail stores and warehouse management), optical near infrared (NIR) for detecting fabric composition and color, e.g., Textile 4 Textile project (Alkazam 2013), and even robotics by integrating haptic and visual sensing, and recognition (CloPeMa 2015). One of the recent initiatives taken has been the FIBERSORT program using NIR spectroscopy, jointly undertaken by Valvan Baling Systems, Metrohm, Worn Again, Fairtex, Reshare, and Circle Economy. Such initiatives not only highlight the process innovation and design in activating circularity but also show how choice of right partners and collaborative networks play a crucial role in bringing together complementary expertise of the actors in several areas. The FIBERSORT project, in this regard, involves strategic partners such as Valvan for sorting machineries, Metrohm for NIR scanning, and Wieland Textiles for secondhand textile processing.

Similar initiatives can be seen in recovering value from end of a product life cycle through chemical recycling, e.g., UK-based Worn Again is developing a chemical textile to textile recycling technology and have partnered with retail brands such as H&M and Kering (Worn Again 2016); Finnish initiative called Relooping Fashion is in the process of closed-loop ecosystem based upon a cellulose dissolution technology to create new clothing out of recycled cellulose and involves various strategic partners along the value chain representing crucial operations such as collection, sorting and recycling, retailing, and distribution (ReLooping Fashion 2016). Pure Waste, a Finnish brand, and partner in the Relooping Fashion initiative, is involved in this process in developing clothes out of industrial wastes (cutting wastes and leftover of the manufacturing process), which is then sorted by color, refibred and finally spun into yarn (Pure waste 2016). Another innovative network-based initiative, the Dutch aWEARness, is taken by various strategic partners such as ecofabric producers, workwear resellers, tracking, and tracing partners to deliver circular workwear to various resellers. Dutch aWEARness works as a circular supply chain content manager by maintaining a database with information about materials, includes a life cycle analysis, a purchasing and inventory management tool and a track and trace system, and in turn receives a service charge (Dutch Awearness 2016).

Compared to recent recycling cellulose projects and initiatives, synthetic fibers have been recycled for a considerable longer time. Returnity is a 100 % recyclable polyester fabric licensed by Dutch aWEARness, used for making workwear and interior-furnishing. By adopting a cradle-to-cradle (C2C) design guideline, the product is said to reduce $CO_2$ impact by 73 %, waste management by 100 %, and water use by 95 % compared with cotton (Perella 2015). Similar products, e.g., Econyl, have been developed by Interface, world's largest modular carpet manufacturer, by reclaiming discarded fishing nets, by entering into a collaborative supply chain partnership with Net-Works enables local residents to collect discarded nets and sell them back into a global supply chain for issuing a second life (Net-works 2016).

In addition, various design strategies address the aspect of planned obsolescence by slowing down the rate of depletion of resources and by optimizing the product life span. Packard (1963) categorized these design strategies for addressing obsolescence of technology, quality, and desire. While *design for reuse/upgrading/easy maintenance/easy replacement* aims at tackling technology obsolescence in a multitude of ways, e.g., designing in module for easy detachment and reattachment, *design for remanufacturing* comprises of design for easy disassembly and assembly. Sinha et al. (2015) highlight a generalized remanufacturing fashion process that is elemental in supporting efficiency and effectiveness of reverse supply chain by meeting scale, speed, and quality issues. Furthermore, strategies for product durability that addresses obsolescence of quality include design for more robust products, while timeless design and degrees of customization ensure extension of product life span by reducing obsolescence of desire (Mont 2008).

Indeed, tight component and product cycles of use and reuse aided by product design create positive opportunities for CE. This further distinguishes CE from linear economy which loses large amounts of embedded materials, energy, and labor. More importantly an integrated systemic view of the whole system is required to design "value from waste."

## 3.3　Archetype 3—Substitute with Renewables and Natural Processes

Currently, the growth-oriented motives of a globalized throwaway society have constantly depleted the Earth's resilience by creating a decrease and imbalance in the natural capital. Rockström and Klum (2012) have highlighted this as the "quadruple squeeze" where 60 % of the key ecosystem resources are utilized in support of human well-being, thus rapidly eroding Earth's resilience potential. Global Footprint Network (2014) measures have shown that the ecological footprint of production and consumption in terms of the Earth's regenerative capacity has increased from little <0.5 in 1960 to 1.5 Earths currently, and by 2050 we will need 2.3 Earths. Such "business as usual" (BAU) would result in severe resource scarcity in the near future. Thus, substitution of fossil fuel-based energy and resource systems by renewable materials is a key to achieve impact reduction on the Earth and environment. Several strategic frameworks and concepts such as Blue economy, Zero Emissions (ZERI), Biomimicry, and The Natural Step are concerned with the potential for creating environmentally benign industrial processes and live within current resource constraints, by making better use of renewable resources (green manufacturing) or drawing inspiration from processes occurring in nature (biomimetics) (Aneja and Pal 2015; Bocken et al. 2014).

The term "green" manufacturing in textiles and apparel can be looked at in two ways (Green Technica 2012):

1. manufacturing of "green" products, particularly those using renewable resources and energy systems or resources having reduced environmental impacts, and
2. "greening" of manufacturing—reducing pollution and waste by minimizing natural resource use, recycling and reusing what was considered waste, and reducing emissions.

In this context, green textile innovation is a broad area of innovation, ranging from replacing chemical dyes with organic/benign dyes in textile production, through to more radical changes such as the emerging field of "green chemistry" that seeks to utilize naturally occurring processes in place of traditional industrial processes. Textile production utilizes diverse chemicals, dyes, and finishes at all stages which are harmful to the environment and health, and its globally dispersed nature makes matter worse by posing a big challenge in stipulating requirements and regulations. LAUNCH Nordic is an initiative taken by Nordic countries to promote use of environmentally friendly materials (based on cleaner manufacturing and green chemistry) in fashion and textiles as one of their focus (LAUNCH Nordic 2016). Under this theme, the initiative focusses on innovations in: (i) optimizing existing technologies and approaches within the closed-loop manufacturing, green chemistry, and recycling/reuse; (ii) reducing the toxic/chemical impact at various points in the value chain, e.g., through use of green chemistry; and (iii) exchanging manufacturing information and data (LAUNCH Nordic 2016). For instance, under this initiative, Novozymes in Denmark is exploring new uses for enzymes in textile production which can reduce a textile's environmental impact and improve the durability of finished products, resulting in reduction of consumption of water, energy, and chemicals (Norden 2015). The initiative also provides scope for new collaborations between LAUNCH global partners and multi-nationals worldwide, e.g., IKEA, Novozymes, and Kvadrat to drive system innovation and help scale sustainable innovations in materials (LAUNCH Nordic 2016). Another aspect of "greening" the manufacturing process is by operating without emissions and waste. Zero Emission Research Initiatives (ZERI) is a research initiative proposed to cascade nutrients, materials, and energy so that our production and consumption systems are designed in such a way (ZERI 2013). On similar lines, Detox campaign (launched in 2011) mobilizes global clothing brands and their suppliers to eliminate toxic water pollution and release of hazardous chemicals from their supply chains and products, and presently includes 19 international fashion companies, e.g., Nike, H&M, and Zara. Overall, the fundamental principles of this campaign are as follows: (i) zero discharge of all hazardous chemicals (wastes and production emissions or later "losses" from the final product), (ii) prevention and precautionary actions toward the elimination of hazardous at source through substitution with sustainable alternatives or even product redesign, and (iii) full transparency by brands and their supply chains and public disclosure of information about hazardous chemicals used and discharged (GreenPeace 2016). Detox Outdoor initiative is a similar open campaign taken by Greenpeace to challenge big outdoor brands to eliminate per- and polyfluorinated chemicals (PFCs) and other hazardous chemicals from their entire production and become Detox Champions (Detox Outdoor 2016).

Further green products can be innovating biological nutrient-based renewable resources aimed at generating a bioeconomy through greater use of renewable biomaterials instead of finite resource elements. In textile production, many alternative sources of raw materials have been explored to replace conventional materials having slow regenerative rate or longer replacement cycle. For example, research is being conducted in the Nordics to look into alternatives for cotton by using chemical wood pulp or recycled biobased textiles (Bio Innovation 2016). Re:newcell is a Swedish innovation start-up which aims at recycling and transforming a high cellulosic portion fabric into recycled dissolving pulp ("re:newcell pulp") which can then be used for commercial textile production (Re:newcell 2016). Further, new green fabrics, materials, and practices are constantly explored to pave the way in TCF manufacturing sector. Sustainable Technology in Nettle Growing (STING) is a collaborative project between the company Camira, Defra (Department for Environment, Food and Rural Affairs), and De Montfort University to develop textile material made out of stinging nettle leaves (Green Product 2014). Many other biocomposites, created through combination of biobased fibers, such as kenaf, hemp, flax, jute, with polymer matrices are also essential in creating biofiber–matrix interface and novel processing.

Beyond, green sustainable or alternative products and processes, learning from and replicating nature to find solutions for answers to various problems related to our quest for new products, processes, or technology, have been for a long time and is termed as "biomimicry" or "biomimetics" (Vincent et al. 2006). Biomimetics incorporate principles that rethink our approach to materials development and processing and hence promote sustainability by reducing ecological footprint (Eadie and Ghosh 2011). Natural systems are inherently energy-efficient and adaptable, hence to be sustainable, textile fibers and products must emulate this feature. Further, the increasing demand for fibers worldwide has driven the use of new and innovative products, to be met largely by using renewable resources and through efficient recycling; however, high number of polymers in fiber structure makes it immensely difficult, at times, to eventually recycle the product. Biomimicry in textiles, in this regard, considers recyclability and aims at reducing the number of polymer types we tend to use.

## 3.4 Archetype 4—Deliver Functionality Rather Than Ownership

This archetype emphasizes servitization and product-service systems (PSSs) (Tukker 2004), wherein functionality and access are valued more over ownership of the product. Services are provided to satisfy users' needs aimed at increasing the active lifetime, single or multiple, of each product. In some cases, the service providers retain the ownership of the product entirely through its lifetime, only offer the usage (performance, functionality) of the product against a fee, and are termed as use-oriented (UO) PSSs. Various rent- and lease-based business models have emerged in TCF sectors in the recent years. VIGGA and Katvig are Scandinavian children's clothing brands based upon such lease-based subscription business model. For instance, VIGGA charges a subscription fee of roughly € 48 per month with a subscription period ranging between 3 and 27 months to receive 20 pieces of clothes. MUD Jeans is a pioneering example of the leasing business model; the member pays a fee of € 25 one time to receive a pair of jeans in return, followed by a payment of € 7.5 on a monthly basis for 12 months. After 12 months, the user receives an email from MUD Jeans prescribing 3 options: "keep them," "switch them," or "send back." While in the first case the user keeps the jeans pair, in the second option he/she pays a switching fee of € 10 for the first month and sends back the old jeans to receive a new pair in return. In the third case, the user sends back the jeans to receive a € 10 voucher to use anytime later (Mud Jeans 2016). Such UO-PSS models have the potential to change the consumption patterns by reducing the need for product ownership, and in addition incentivize the manufacturers and brands in developing products that last longer through various "design for …" approaches, e.g., "design for upgradability" and "design for reparability."

In the renting model, many clothing brands have launched this new concept, such as Uniforms for the Dedicated (under the concept "The Collection Library"), Fillipa K ("Make it Last"), Rent-a-Plagg ("Are-360"), among others allowing customers to rent key pieces from current collections—primarily special occasion wear such as suits, dresses, and accessories. Several online retailers, like Rent the Runway, Le Tote, etc. have also ventured into such rental schemes, by renting out designer labels under various rental schemes (fees and return periods) and additional services (e.g., free drop-off, style, and mix-match suggestions).

In contrast, many brands have launched services facilitating product life extension, through repair and maintenance services and/or by selling a durable product through warranty. Nudie Jeans, for example, is a Scandinavian denim brand which offers its customers free in-store repair services which contributes largely to the sustainability image of the brand. For consumers who are unable to visit the store are offered free DIY repair kits. In these PSSs, the ownership lies with the customer; however, such servitization schemes aim at extending the product lifetime and durability and are called product-oriented (PO) PSSs.

In connection, collaborative consumption is also an important movement changing the consumption landscape based upon the idea of sharing and collaborating to meet certain needs, products, and services. In the TCF industries, collaborative or sharing business models are mushrooming in the recent years, e.g., online marketplaces like ThredUp, which allows peer-to-peer selling of secondhand clothing, swapping platforms like Swapstyle, which enables people to swap fashion worldwide, and local Swishing parties where people share clothes from their own wardrobes. In this regard, clothing or fashion libraries have emerged largely in the scene, which are subscription-based service that allows people to share wardrobes (Pedersen and Netter 2015).

Overall, such servitized SBMs offer and build a unique relationship with the consumer/user which enhances the brand loyalty and consumer insight. Renting/leasing activities work best either for durable and high-quality products (e.g., workwear, denims) or for seasonal products (e.g., kids wear, maternity wear) mainly because consumers/users find it cheaper than buying while the companies can rent/lease out the product multiple times against a fee (Circle Economy 2015). Repairing services, on the other hand, are particularly attractive for garments that easily experience wear and tear and are expensive (e.g., outdoor gear and jeans). Sharing and swapping activities are particularly favorable for clothes which have crossed the end-of-use period during a single lifetime, either permanently or temporarily, yet holds a potential for usage in subsequent lives.

These SBMs do not require immediate attention toward design strategies, however, in long-term calls for product design for durability, and new partnerships in the network for supporting the value recovery processes, e.g., repairing and washing or end-of-life recycling. VIGGA, for instance, collaborates with an external recycling facility where the baby clothes no longer fit for circulation (roughly after 82 weeks) are transferred to recycle the fibers.

The fundamental change brought but various SBMs representing this archetype are in terms of reducing the material throughput and resource consumption patterns in the industrial system, and enhancing durability and longevity by increasing the active usage time of the product through reparability, upgradability, etc. (Reim et al. 2015). However, it is uncertain whether PSSs are truly ecoefficient—renting or leasing may not tend to displace purchase of newer products—unless being considered simultaneously with other archetypes, for example, "creating value for waste," as in case of sharing of secondhand clothes or recycling (Bocken et al. 2014).

## 3.5  Archetype 5—Adopt a Stewardship Role

Along this archetype, companies engage in undertaking corporate stewardship, meaning a role to ensure positive impact on the health and well-being of the stakeholders (society and environment).

Typically, the consumer pays a price premium to fund the benefits of such stewardship role executed by manufacturer or retailer along the supply chain. Upstream stewardship activities incorporate choice of supplier and production system to deliver ethical and sustainable business practices, such as fair labor wages, corporate citizenship and community development, and environmental protection. In the TCF industries, brands and retailers increasingly show upstream stewardship by reconfiguring their networks to ensure accredited suppliers and production processes.

One exemplar of exercising stewardship is through third-party certification schemes which are availed to show compliance to set standards and be transparent to the consumers about the production and supply chain operations. Numerous ecolabels are available in the market (e.g., EU Ecolabel, Nordic Ecolabel, Swedish Bra Miljöval, Bluesign, GOTS (for organic textiles), Ökotex (for harmful chemicals, etc.) aimed at facilitating, supporting, or monitoring sustainable practices in sourcing and production (Pal 2014); in addition, individual companies have developed their own labels, e.g., H&M Conscious (Norden 2015).

From the process level, such standards and ecolabels like Bluesign® system drives toward sustainable textile production through management of "Input Stream" from raw materials to chemical components, to eliminate harmful substances from the beginning of the manufacturing process and set and control standards for an environmentally friendly, safer production (Pal 2014). This ensures that the final textile product meets very stringent consumer safety requirements worldwide, but also provides confidence to the consumer to acquire a sustainable product. On the one hand, Bocken et al. (2014) highlight that such stewardship strategies can generate greater brand value and potential for premium pricing and Norden (2015) states that this is not yet certain in the TCF sector.

The Higgs Index is a widely accepted tool developed by sustainable apparel coalition (SAC) that measures the environmental and social impact of fashion supply chains. Fashion brands, such as H&M, Patagonia, Adidas, Asics, Coca-Cola Company, New Balance, Nike, and Puma, as a member, can self-assess their sustainability efforts throughout a product's entire life cycle and design their scoring scale to communicate a product's sustainability impact to consumers and other stakeholders (Pal 2014; Sustainable Apparel Coalition 2016). SAC through its alliance actively strives to measure and benchmark sustainability performance and achieve environmental and social transparency that consumers are starting to demand, hence advocate responsibility.

In connection, public green procurement of textiles and apparel is yet another initiative taken at the EU level to determine and guarantee the import of higher amount of resource-efficient products those meeting certain quality parameters based upon the European standards that are commonly used throughout the industry. Currently, this is set as a voluntary guideline by EU to procure 50 % based upon green procurement criteria but is aimed to be a directive in future (Norden 2015). Similarly, a Nordic guideline and cooperation on green procurement is set to be launched in 2016 to stipulate criteria for ecolabelling, environmental management, reuse, and the durability of textiles. This considerably demands network configuration and collaboration, as for example, the initiative will promote the joint initiative between the ecolabelling organizations and the industry to develop marketing ideas, collaborative relationship with manufacturers and brands who aim to have the ecolabel on their products, and also with producers and subcontractors.

Along the supply chain, such certifications, standards, and ecolabels are supposed to enhance visibility and radical transparency about environmental/social impacts. For example, Nike through Fair Labor Organization (FLA), an NGO, openly shares the results of the audits of its suppliers including disclosure of its factory details for maintaining degrees of transparency. Further, it has developed a Sourcing and Manufacturing Sustainability Index for assessing the strategic performance of its sourcing and publishes it in yearly Corporate Social Responsibility (CSR) report. Similarly, Nudie Jeans through its Web site shares the complete production guide which includes details of its suppliers, locations, audit reports, production capacities, and conditions. In addition, it discloses information about CSR, by publishing the code of conduct, social report, brand performance check report, etc., publicly through its Webpage. Many other fashion brands and designers have started to get associated to ethical and fair trade fashion; for example, the Ethical Fashion Initiative is a project initiated by the International Trade Centre aimed at responsible fashion industry by linking high-end fashion designers such as Vivienne Westwood, Karen Walker, Marni, and Stella McCartney with marginalized artisans in a number of African countries (Smith and Newman 2014). This ensures a number of benefits: Workers earn a living wage, are offered dignified working conditions, and minimize impact on the environment (Ethical Fashion Initiative 2016).

From the retailers' frontier, another way to show stewardship is through "choice editing," which refers to how brands and retailers cut out environmentally offensive products and introduce real sustainable choices on the shelves for their mainstream consumers (Sustainable Consumption Roundtable 2006). Various strategies followed are, such as (i) removing products from commercial consideration, or (ii) making products expensive to use, which directly impacts the consumption and aims to only provide sustainable products in the market. Choice editing also incorporates editing out or replacing product components, processes, and business models in partnership with other actors in society such as policy-makers (WBCSD 2008). Adidas Group, for example, takes a proactive step to control its globally dispersed supply chain and actively edit the environmental impact, by supporting its suppliers in order to reduce their environmental impact by developing training materials, technical guidelines, and workshop tailored to each supplier's special need (WBCSD 2008). Teijin, on the other hand, has shifted its focus toward editing of business process model by developing a recycling system for polyester (under its ECO CIRCLE program) which reduces energy and resource use, $CO_2$ emission, and waste. Such a closed-loop recycling business is established in partnership with a global network of companies that collect polyester garments for recycling and advocate the development and marketing of products containing recycled polyester (Teijin 2016; WBCSD 2008).

Additionally, clothing brands and retailers undertake downstream stewardship by implementing extended responsibility even after selling away the product. Unlike many sectors, e.g., waste electrical and electronic equipment (WEEE) where such responsibilities are mandatory and are set under environmental producer responsibility (EPR) directives, in textile and clothing, no country except France has a mandatory EPR scheme. However, to ensure greater responsibility from the brands and retailers importing garments, a proposal is drawn up in some Nordic countries, to ensure some sort of "polluter pays principle" commitment (Ekvall et al. 2014). This includes planning for and, if necessary, paying for the recycling or disposal of the product at the end of its useful life. This may be achieved partly by redesigning products to use fewer harmful substances, to be more durable, reusable, and recyclable, and to make products from recycled materials. For retailers and consumers, this means taking an active role in ensuring the proper disposal or recycling of an end-of-life product. Implementation of such EPR schemes in TCF sectors are mend to implement an extensive system where the importers execute responsibility in take-back schemes for collection of used/waste textiles and clothes. Such stewardship role through exercising EPR schemes, mandatory or voluntary for showing individual and collective responsibility, is expected to increase the potential for recycling and reuse of textiles and also generate scope for new SBMs and have been increasingly prioritized in EU and Nordic Region (Ekvall et al. 2014). Carefully designed EPR schemes can provide various incentives and upstream effects to the stakeholders, for example, rebates on producer participation fees, reductions in the use of certain chemicals during the production of textiles, designing for a longer life, and avoiding fiber mixes to allow easier recycling at end of life (Ekvall et al. 2014; Lindhqvist 1992).

## 3.6   Archetype 6—Encourage Sufficiency

Encouraging sufficiency actively seeks to reduce or moderate rate and volume of consumption, thus leading to a fundamental change in the Western economic model based on market economy, throwaway paradigm, and planned obsolescence (Bocken and Short 2016; Jackson 2009; Mont 2008).

Drivers of such business model innovation for sustainability lie in "curbing demand through education and consumer engagement, making products that last longer and avoiding built-in obsolescence, focusing on satisfying 'needs,' rather than promoting 'wants' and fast fashion, conscious sales and marketing techniques, new revenue models, or innovative technology solutions" (Bocken and Short 2016).

Bocken and Short (2016) further highlights that such sufficiency-based business model innovations adopt varied value creating logic, as specified above and together with other SBM archetypes can be seen as a holistic strategy to reduce over-consumption and production leading to a sustainable future. This certainly goes beyond the commitments of earlier stated ecoefficiency-based business models (e.g., saving energy and materials, green economy, and circular economy) which may facilitate rebound-effects where efficiency gains lead to more consumption (Bocken et al. 2014; Druckman et al. 2011).

Sufficiency being embedded through the holistic design of the business model across all aspects of business focusses in many ways toward product design changes to enhance durability, reparability, and longevity (Bocken et al. 2014; De Groene Zaak and Ethica 2015). Many luxury brands, for instance, offer timeless design and high degrees of craftsmanship and promote "slow fashion" which ensures diverse practices of supporting local manufacturing, durable or timeless product designs, reuse activities, and slow consumption (Fletcher 2010; Fletcher and Grose 2012). Apart from paying a high premium price which in most cases reflects upon high degrees of artisanal production and high-grade materials, consumers also develop a strong emotional attachment with the product, thus instigating longevity in usage. This has the potential to eschew fast fashion trends. Other luxury and slow fashion brands, apart from contributing to sustainability along other archetypes, have been able to generate their timeless design profile vis-à-vis improved ethical practices and improving traceability to embed sufficiency. In the recent year, particular attention has been shown in research frontier toward investigating sustainability and luxury together, considering their striking similarity. The Center for Studies on Sustainable Luxury (Center for Study of Sustainable Luxury 2016) in Argentina is one such research group that explores how luxury brands could represent the greatest positive contribution to people and planet by creating "deeper," "more authentic" meaning of "luxury" to motivate social and environmental performance.

From the social perspective, many luxury and slow fashion brands have embraced ethical practices to communicate sufficiency. People tree, for instance, is a UK-based slow fashion fair trade brand working closely with textile women artisan groups from Bangladesh to help them meet environmental standards, promote ethical sourcing and social value creation, vis-à-vis ethical consumption of hand woven and natural dyed products—thus eliminating built-in obsolescence. As mentioned above under Archetype 5, the Ethical Fashion Initiative promotes responsibility along the fashion value chain by ensuring a number of benefits to the workers, e.g., earning a living wage, dignified working conditions and minimum impact on the environment and on the other hand communicates this added value and quality brought by artisanal production (Ethical Fashion Initiative 2016) to promote "buy less for more" concept. Larger companies, such as Patagonia and Nudie Jeans, take much more diverse efforts in communicating sufficiency through their business. Patagonia, for instance, engages with activities for moderating sales by organizing manipulative consumer marketing campaigns, no sales incentives, choice editing, etc. One way of communicating their unique business model has been via their "don't buy this jacket" campaign intending to encourage people to consider the effect of consumerism on the environment and purchase only what they need (Ekvall et al. 2014). In addition, Patagonia's "Common Threads Initiative" supports "reduce, repair, reuse, and recycle" activities emphasizing the waste hierarchy. Partnerships with iFixit teaches customers how to repair their gear to increase the useful life of products under the "Buy Less, Repair More" campaign (iFixit 2015), while that with online marketplace e-Bay supports and encourages reuse of secondhand Patagonia clothing once unrepairable. This promotes diverse value creating logic, while follow-on repair services yields long-term customer relationships and trust, leading to loyalty and reputational benefits, collaboration with online marketplace renders a strong resale value of used product, thus encouraging customers pay premium price. As mentioned above, Nudie Jeans similarly encourages sufficiency along its eco-cycle initiative (Nudie 2015); Nudie Repair shops help its customers to repair worn out jeans free of charge or give back the jeans totally, which is then washed and repaired and put back in the shop as a secondhand item for sales (Pal 2016). Likewise, smaller designer-led initiatives have also emerged in the scene in the recent years which promote sufficiency by offering value regaining services. Both online and physical platforms and initiatives have been launched which offer redesign or repairing solutions as pay-per services or completely free of cost, or educational awareness through lessons. By embodying multiple mechanisms within the business model, these SBMs render sufficiency along with. For example, Dream and Awake is a small designer-led initiative in Sweden that collects or buys old vintage clothes from the market and redesigns, photographs, and finally sells them through mobile studios. It also involves with organizing redesign workshops with wearers in providing designs, facilities, and equipment to help them redesign their own clothes (Pal 2016). This way, it drives redesign service-oriented initiative to encourage people to mend their garments and extend their use value thus potentially slow down the replacement cycle.

In general, the argument whether reuse and sales of secondhand clothes contribute positively toward sufficiency is dichotomous. Bocken and Short (2016) highlight that sales of secondhand goods may incentivize owners to take more care of the products to ensure higher secondhand value; however, Ekvall et al. (2014) in contrast claim that the displacement effect may be negligible considering that most of the time the secondhand garments donated to charities are not priced or when disposed via take-back schemes are valued by a reward mechanism which may instigate newer purchases. However, some luxury secondhand shops (Affordable Luxury, Beyond Retro) unlike the other resale arrangements run for profit, where individuals can leave their garments for sale, and then split the profit with the consignment shop. This can increase levels of reuse by making use of the large amounts of clothes which are hanging unused in wardrobes by giving them a value.

### 3.7 Archetype 7—Repurpose the Business for Society/Environment

Social enterprises and social businesses are those which prioritize social and environmental benefits rather than the economic profits, by incorporating a "social profit equation" into the business model (Bocken et al. 2014; Yunus et al. 2010). Social business was defined by Nobel laureate Professor Muhammad Yunus as "the new kind of capitalism that serves humanity's most pressing needs" (Yunus 2007).

Between for-profit and nonprofit business approaches, social enterprises are driven by the fundamental purpose of delivering social and environmental benefits primarily, however, not overlooking the economic aspect of it. The benefits are mainly in terms of creating self-sustaining business operations by either: (i) focusing on businesses dealing with social objectives only, or (ii) by taking up any profitable business (so long as it is owned by the poor and the disadvantaged) who can gain through receiving direct dividends or by some indirect benefits (Yunus 2007).

On the one hand, various charity organizations fundamentally work along the nonprofit concept in the secondhand clothing sector by solely committing to collection of used clothes from Western countries and arranging donations for humanitarian purposes in under-developed nations and for emergency responses in crises regions. Red Cross, Salvation Army, and Oxfam are the large charity organizations working globally to take part in such activities.

However, these organizations are increasingly being noted to run a hybrid business model, whereby two business entities coexist, one operating as a traditional for-profit business, but using part of the profit stream to finance a second not-for-profit enterprise (Bocken et al. 2014). For instance, in the secondhand clothing sector, Salvation Army has developed its trading arm called Salvation Army Trading Company Limited (SATCoL) which processes the donated clothes and exports it to buyers in Eastern Europe; Oxfam International (an exported-oriented subsidiary of Oxfam) on the other hand exports 50 % of its total collection to West Africa to traders in those markets, thus creating business opportunities (Brooks 2013). Similarly, Fretex is company managing textile collection operation in Norway and is owned by The Salvation Army. Fretex International, jointly owned by Fretex, Norway, and Myrona, Sweden, is the wing that generates business opportunities by exporting part of the collected garments to Africa and Asia. Sometimes such operating models are classified by charities as social businesses, defending the profit-making export orientation, by organizing value-added activities locally in poorer countries. Oxfam, for instance, has established a pilot local processing enterprise in Senegal called *Frip Ethique* aimed at generating local employment (Brooks 2013). However, there are a lot of rising criticisms against this trade as underlying business motives and transactions are often not publicized and are concealed back-stage in contrast to the foregrounded charitable acts of donation (Brooks 2013). On the other hand, there are small fashion social enterprises with diverse social fundaments, such as supporting recovering addicts, refugees, and aging. The North Circular is a UK-based producer of luxury knitwear that uses a local network of talented home knitters (mostly aging woman) to mobilize localized production (The North Circular 2016), while Who Made Your Pants?, an UK-based Lingerie brand, employs and supports women refugees from the Southampton area. Set up as workers' cooperative, it is funded by small grants and revenue and any profit made is returned to the business and any leftover is shared between the members and democratically agreed good causes (Reddy 2014).

On a systems level, such social and/or hybrid business models integrate business with varied stakeholders through participatory approaches, which may include nontraditional approaches (e.g., collaboration between for-profit and nonprofit organizations or with the local community) or new organizational designs (e.g., hybrid structures).

Interorganizational collaboration in this regard has emerged very strongly to shift from the fundamentals of ego-centric business models to more altruistic collaborative business models (CBMs). CBMs in this context refer to a value creating system or network where multiple organizations that might differ in type (industry, public research, nonprofit), their position in the value chain (manufacturing, service, etc.), and industry and work together to create and capture value at the systemic level, more than the value created and captured for each stakeholder (Breuer and Lüdeke-Freund 2014; Rohrbeck et al. 2013). In a smaller level, the charity organizations have collaborated with fashion retailers to organize collection and take-back schemes, with logistics providers for easy in situ collections and distributions, and sometimes with reprocessing organizations like laundries. Fretex for instance cooperate with major retailers like Lindex in Sweden for collection, and also with Norwegian Postal Service to provide the possibility to consumers to discard their old clothes in specially designed bags. Major fashion retailers and brands, for example, have either organized secondhand sales business on their own (especially brands, e.g., Fillipa K, Boomerang) or in collaboration with global sorters like I: Collect, as done by H&M. Any financial profit made through such initiative is utilized for social and charitable activities. For example, H&M donates 0.02 euro to a local charity organization for each kilogram of clothes.

Another aspect of such socially driven business enterprises and cooperatives is the aspect of localness and upliftment of local community. Several initiatives have already originated in Asia and Africa aimed at empowering women and driving community forward. As mentioned earlier, SEW and Sidai Designs are some exemplar cases of fashion social enterprises as well, providing sustainable employment to Tanzanian women by selling their designs and/or ethically manufactured collections through several retailers in the UK, USA, and Australia (Smith and Newman 2014). While on the one end, ethical fashion as mentioned above creates opportunity for established fashion houses to execute corporate stewardship and on the other provides scope for development of a local manufacturing industry in poorer countries by upholding marginalized artisanship.

## 3.8  Archetype 8—Develop Scale-up Solutions

Beyond the traditional view on scaling-up businesses through product-process optimization and standardization, widespread implementation of SBMs requires alternative thinking on scalability. Currently, even though various SBMs have emerged in the scene with positive impact, they are mostly small scale or niche. As in other industries, TCF industries face the same challenge in terms of scaling-up sustainable business models and ideas. Although large firms (manufacturers and retailers), e.g., H&M, G-Star, are in the forefront of driving sustainability, these initiatives are still in their infancy compared to mainstream business models. For instance, H&M in cooperation with I:Co has entered into a take-back arrangement to collect used clothes from consumers, but the extent is of the order of 7600 tons (in 2014), compared to the amount of clothing sold worldwide every year ($\sim 150$ million tons) or disposed-off. New start-ups and small businesses undertaking the more radical innovations are also at a niche or small scale. Thus, the need to scale-up SBMs to a global mainstream level is of utmost importance. Bocken et al. (2014) provide many exemplary cases underpinning this archetype, out of which many can be seen to be emerging strongly in the TCF sector, viz. peer-to-peer (P2P), crowd-based, platform-based, open innovation, cocreation, and digitalization.

Collaborative business models in this context provide a scope for rapid scaling up and include various exemplar cases, such as crowd-sourcing/crowdfunding and open innovation. Collaborative networked organizations (CNOs) tend to open up their boundaries to provide opportunities for other businesses to thrive (Saebi and Foss 2014). Such collaborations can be the basis for cocreating products or for crowd-sourcing ideas. Such cocreation spans from passive involvement of users, as in niche e-commerce for pretailing, to mass customization to active user involvement, where the users are also involved as "inventors." Threadless is an online community of artists and an e-commerce Web site based in Chicago, which involves such "inventive" users by putting designs created and chosen by the online user community to a public vote. A week later, the top-scoring designs are reviewed by the staff to put into production each week and are finally sold worldwide. Designers whose work is printed receive 20 % royalties based on the net profits and $250 in Threadless gift cards. Such crowd-sourced business model attains scalability through its innovative network and active engagement with the user community. Many similar online crowdfunding start-ups, such as Out of X, Carte Blanche, and Cut on Your Bias, are aimed at demand management by the use of system, software, and other communication channels.

In a value network context, such crowd-sourced models and CNOs can be open platforms assimilated through information technology (IT) tools to integrate producers, suppliers, and customers along various activities (Liu et al. 2014; Romero and Molina 2011). Such multi-sided platforms act as an intermediary between sellers and buyers as in case of online marketplaces, thus connecting multiple stakeholders, third parties, and consumers. Open Garments is an EU-funded research initiative based upon such open innovation concept where: (i) a virtual user consumer community designs, configures, orders, publishes, shares, and even sell individual garments; (ii) an open manufacturing-based flexible network of production units (mainly microenterprises) produces customized physical goods; and (iii) a knowledge-based manufacturing service provider (MSP), which is the open platform, acts as a service provider in between (Open Garments 2009). Such open innovation platforms, apart from redefining the collaborative format (where users are designers), also attempt to change the consumption pattern by radically influencing the production model. Mina et al. (2014) explain that the service innovators as a central component of such platform of interactions (e.g., Web-based collaboration platforms) support both upstream and downstream activities and are responsible for building the infrastructure to connect other stakeholders, third parties, and consumers.

On the other hand, such platforms can be also at the P2P level as in case of Web-based swapping platforms, e.g., Kleiderkreisel offers online platform-based sales and purchase services for used clothes. It further offers its members social networking and communications to promote collaborative consumption and swapping through the creation of a platform-based community. Such platforms for P2P interaction aim at either value cocreation or collaborative consumption.

## 4 Sustainable Design Elements

The core notion of sustainable design is uphold by systems thinking (Evans et al. 2009), aimed at creating a sustaining industrial system based upon product, process, and facility design, for enhancing the well-being of nature and culture while generating economic value (McDonough and Braungart 2002). The traditional perspective of sustainable design encompasses the intention to "eliminate negative environmental impact" through manifestations of renewable resources, ecoefficiency, etc., thus impacting the environment minimally. In-line with Tischner and Charter (2001), this encompasses the aspects of repair and refine, thus emphasizing the notion of "[…] modifications to existing products, with some movement toward increasing the eco-efficiency of existing products."

However, beyond the "elimination of negative environmental impact," sustainable design must create meaningful innovations that can create a dynamic balance between economy and society, intended to generate long-term relationships between user and object/service. Tischner and Charter (2001) propose that this addresses a redesign approach, especially in the use of new technologies and materials to reduce the environmental impact of products. Managing such innovations from a value network perspective fosters cross-industry partnerships such that different actors can cultivate their strengths such as regional presence, customer and market access, legislative power, infrastructure competencies, and know-how (Breuer and Lüdeke-Freund 2014; Calia et al. 2007). Niinimäki and Hassi (2011) further highlight the need to focus on both production and consumption, as a system in whole, for creating sustainable development. Hence, this requires *rethinking*, or a radical change in mind-set, and it can offer breakthroughs for new lifestyles, the ways of living and doing things, as well as approaches to fulfill consumer needs in a more sustainable manner. This approach leads to strategic innovations for generating SBMs and is underpinned by design for sustainability in any industrial system along 5 key design elements:

1. Product design,
2. Process design,
3. Value network design,
4. Relational design, and
5. Design of "new" consumption pattern.

Table 1 shows a breakdown of the SBMs discussed under each archetype (in Sect. 3) into their corresponding design elements.

Nonrepresentation under any particular design element for an archetype does not mean that those archetypes are devoid of or not underpinned by a suitable design strategy, but merely shows that those elements are not the key aspects underlying the design of the archetype. Further, these design elements (and strategies) under an archetype are not mutually exhaustive and in most cases are integrally related to other design elements (and strategies), thus adopting a systemic perspective. In addition, some of the archetypes are overlapping (e.g., Archetype 2: "Creating value from waste," and Archetype 4: "Deliver functionality rather than ownership"); hence, some of the exemplar SBM cases along with their underlying design elements (and strategies) cannot be uniquely categorized under one archetype.

**Table 1** SBM archetypes along 5 design elements

| Archetypes (*major initiatives*) | Design elements | | | | |
|---|---|---|---|---|---|
| | Product | Process | Value network | Relational | Consumption pattern |
| **Archetype 1**<br>*Company-driven initiatives (Clothing Continental ), Materialise* | ✓<br>• Ecoproduct design (dematerialization or multi-functionality)<br>• Circular/regenerative design (e.g., mono-materiality, modular design)<br>• 3D printing | ✓<br>• Lean for waste reduction (higher efficiency)<br>• Cleaner Production (use of renewable sources)<br>• Additive manufacturing | ✓<br>• Collaboration between designers, engineers, and developers | | |
| **Archetype 2**<br>*FIBERSORT, Relooping Fashion, Worn Again, Pure Waste, Dutch aWEARness, Returnity, Econyl, Net-Works* | ✓<br>• Design for longevity, durability, ease of maintenance and repair, and if required upgradability<br>• Redesign-make (e.g., easy disassembly and assembly)<br>• Cradle-to-cradle<br>• Robust design<br>• Timeless design<br>• Customized design | ✓<br>• Repairing, upgrading, remanufacturing<br>• Sorting techniques and fiber separation processes<br>• Recycling process and technology | ✓<br>• Reverse logistics (e.g., for take-back schemes)<br>• Collaborative networks based on complementary process expertise (for "closing the loop" activities/operations)<br>• Circular supply chain system and information management<br>• Inclusive business networks | ✓<br>• P2P interaction<br>• Long term with clients/users through access to service<br>• Ethical supply and CSR | ✓<br>• Sharing (use oriented)<br>• Collaborative (P2P)<br>• Product as service<br>• Reduced obsolescence of desire and extended product life span |
| **Archetype 3**<br>*LAUNCH Nordic, ZERI, Detox campaigns, Bio Innovation, Re: newcell, STING* | ✓<br>• Green product design (e.g., green chemistry, biobased)<br>• Regenerative products<br>• Biomimetics | ✓<br>• Green and benign manufacturing<br>• Zero waste and emission processes | ✓<br>• Private–public collaborations for complementary process expertise | ✓<br>• Transparency through manufacturing information sharing<br>• Collaborations between global partners and multi-nationals | |

(continued)

**Table 1** (continued)

| Archetypes (major initiatives) | Design elements | | Value network | Relational | Consumption pattern |
|---|---|---|---|---|---|
| | Product | Process | | | |
| **Archetype 4**<br>Company-driven initiatives (*MUD Jeans, Nudie, VIGGA, Swapping platforms, online marketplaces, renting and leasing*) | ✓<br>• Product-service instead of product<br>• Design for durability, upgradability, reparability, … | ✓<br>• Repairing, upgrading, and maintenance | ✓<br>• Collaborative network of partners for supporting value recovery processes | ✓<br>• Long term with clients/users through access to service<br>• P2P interaction | ✓<br>• Sharing (use oriented)<br>• Collaborative (P2P)<br>• Product as service<br>• Reduced obsolescence of desire and extended product life span |
| **Archetype 5**<br>Sustainable Apparel Coalition, Ecolabel initiatives, Company-driven initiatives, The Ethical Fashion Initiative, ECO CIRCLE | ✓<br>• Circular/regenerative design | ✓<br>• Input stream management<br>• Green public procurement<br>• Process editing | ✓<br>• Reconfigured network of accredited suppliers and production processes<br>• Joint initiatives (e.g., between industry and certifiers)<br>• Inclusive business networks<br>• Network of partners for supporting value recovery processes | ✓<br>• Upstream stewardship through supplier compliance and governance<br>• Upstream stewardship through transparent communication<br>• Collaborative relationship with manufacturers and brands<br>• Downstream stewardship through EPR | ✓<br>• Choice editing |

(continued)

**Table 1** (continued)

| Archetypes (*major initiatives*) | Design elements | | | | |
|---|---|---|---|---|---|
| | Product | Process | Value network | Relational | Consumption pattern |
| **Archetype 6**<br>*Company-driven initiatives (e.g., Patagonia, Nudie, People Tree, luxury and slow fashion brands), Center for Studies on Sustainable Luxury, The Ethical Fashion Initiative* | ✓<br>• Design for durability, reparability and longevity<br>• Slow fashion | ✓<br>• Local or artisanal production<br>• Repairing, upgrading, and maintenance | ✓<br>• Local value network<br>• Collaborative networks (for "closing the loop" activities/operations) | ✓<br>• Consumer education and engagement<br>• Emotional attachment with product<br>• Transparency of ethical practices and traceability | ✓<br>• Satisfying "needs" rather than promoting "wants"<br>• Reduced obsolescence of desire and extended product life span |
| **Archetype 7**<br>*Charity-driven hybrid businesses, Social business-driven, Yunus Centre, Fashion social enterprises* | | | ✓<br>• Inclusive business networks<br>• Promoting local value networks<br>• Cooperatives<br>• Collaborative networks between industry, public research, nonprofit | ✓<br>• Local community development | ✓<br>• Extended product life span (after end of use/end of life)<br>• Ethical consumption paying premium |
| **Archetype 8**<br>*Company-driven initiatives (e.g., Threadless, Cut on Your Bias), online marketplaces, Open Garments, Swapping platforms* | ✓<br>• Product cocreation, Crowd-sourced design | | ✓<br>• Collaborative open and flexible networks<br>• IT-enabled integration of producers, suppliers and customers | ✓<br>• Passive to active user involvement<br>• P2P interaction<br>• Social networking | ✓<br>• Cocreation<br>• Collaborative (P2P) |

# 5 Concluding Remarks

The literature and practice of innovations for sustainability is gaining increasing momentum amidst the ardent need for designing a sustainable society and economy. The TCF industries being the most resource draining and socially unethical in nature call for increased attention. Despite this, as Bocken et al. (2014) highlighted, both knowledge and industrial practices are fragmented and thus need subsequent categorization to make the ongoing and future initiatives more streamlined and impactful. This chapter uses the archetypes proposed in Bocken et al. (2014) (aimed to categorize and explain SBMs) as the starting point, in context to TCF industries, and goes further to identify the key design elements underpinning these SBMs (along the archetypes). The knowledge of these key design elements, upholding a system thinking approach, will further assist both research and industry to strategically develop their SBMs by (i) identifying which design elements need further intervention or modification, (ii) how to enhance sustainability impacts by combining design strategies from other archetypes, and (iii) how to enable a business model change subsequently to derisk the innovation process.

# References

Alkazam (2013) Textile science & engineering automated sorting technology from t4t can help improve recovery and efficiency. J Text Sci Eng 3(3):3–5

Aneja A, Pal R (2015) Textile sustainability: major frameworks and strategic solutions. In: Muthu SS (ed) Handbook of sustainable apparel production. CRC Press, Boca Raton, pp 289–306

Bio Innovation (2016) Etablera närodlad textil i Sverige [Establish locally produced textiles in Sweden]. http://www.bioinnovation.se/projekt/narodlad-textil-i-sverige/. Accessed Apr 2016

Bocken NMP, Short SW (2016) Towards a sufficiency-driven business model: experiences and opportunities. Environ Innov Soc Transitions 18:41–61

Bocken NMP, Short SW, Rana P, Evans S (2013) A value mapping tool for sustainable business modelling. Corp Gov 13(5):482–497

Bocken NMP, Short SW, Rana P, Evans S (2014) A literature and practice review to develop sustainable business model archetypes. J Clean Prod 65:42–56

Boons F, Lüdeke-Freund F (2013) Business models for sustainable innovation: state-of-the-art and steps towards a research agenda. J Clean Prod 45:9–19

Breuer H, Lüdeke-Freund F (2014) Normative innovation for sustainable business models in value networks. Paper presented at the XXV ISPIM conference—Innovation for Sustainable Economy & Society, Dublin, Ireland

Brooks A (2013) Stretching global production networks: the international second-hand clothing trade. Geoforum 44:10–22

Brundtland Report (1987) World commission on environment and development: our common future. Oxford University Press, UK

Calia R, Guerrini F, Moura G (2007) Innovation networks: from technological development to business model reconfiguration. Technovation 27(8):426–432

Center for Study of Sustainable Luxury (2016) http://lujosustentable-eng.org.ar/. Accessed Apr 2016

Circle Economy (2015) Service-based business models and circular strategies for textiles. In: SITRA (ed) Amsterdam

CloPeMa (2015) Clothes perception and manipulation. http://www.clopema.eu/. Accessed Nov 2015

DEFRA (2008) Sustainable clothing roadmap briefing note. Dec 2007

De Groene Zaak, Ethica (2015) Boosting circular design for a circular economy. In E. C. K. P. i. Practice (ed)

Detox Outdoor (2016) http://detox-outdoor.org/en/campaign/. Accessed Apr 2016

Druckman A, Chitnis M, Sorrell S, Jackson T (2011) Missing carbon reductions? Exploring rebound and backfire effects in UK households. Energy Policy 39:3572–3581

Dutch Awearness (2016) http://dutchawearness.com/chain-management/. Accessed Apr 2016

Eadie L, Ghosh TK (2011) Biomimicry in textiles: past, present and potential. An overview. J R Soc Interface 8(59):761–775

Ekvall T, Watson D, Kiørboe N, Palm D, Tekie H, Harris H, … Lyng K-A (2014) EPR systems and new business models: reuse and recycling of textiles in the Nordic region. In Norden (ed) TemaNord, ISSN 0908-6692; 2014:539. Copenhagen

Ellen McArthur Foundation (2013) Towards the circular economy. Economic and business rationale for an accelerated transition, vol 1. Ellen MacArthur Foundation

Ethical Fashion Initiative (2016) http://ethicalfashioninitiative.org/. Accessed Apr 2016

Evans S, Bergendahl M, Gregory M, Ryan C (2009) Towards a sustainable industrial system. With recommendations for education, research, industry and policy. http://www.ifm.eng.cam.ac.uk/uploads/Resources/Reports/industrial_sustainability_report.pdf. Accessed Apr 2016

Fletcher K (2008) Sustainable fashion and textiles: design journeys. Earthscan, London

Fletcher K (2010) Slow fashion: an invitation for systems change. Fashion Pract 2(2):259–266

Fletcher K, Grose L (2012) Fashion and sustainability: design for change. Laurence King Publishers, London

Fuad-Luke A (2009) Design activism: beautiful strangeness for a sustainable world. Earthscan, London

Gardetti MA (2016) Cubreme® and sustainable value creation: a diagnosis. In: Muthu SS, Gardetti MA (eds) Green fashion, vol 1. Springer, Singapore, pp 1–23

Global Footprint Network (2014) http://www.footprintnetwork.org/en/index.php/GFN/page/world_footprint/. February

Green Product A (2014) Nettles in textiles. https://www.gp-award.com/en/produkte/nettles-textiles. Accessed Apr 2016

Green Technica (2012) Renewable energy & clean technology: keys to a revitalization of US manufacturing & job creation. http://cleantechnica.com/2012/04/15/green-manufacturing/. Accessed Apr 2016)

GreenPeace (2016) The detox campaign. http://www.greenpeace.org/international/en/campaigns/detox/water/detox/intro/. Accessed Apr 2016

Gwilt A (2014) A practical guide to sustainable fashion. Bloomsbury Publishing, London

Hart SL, Milstein MB (2003) Creating sustainable value. Acad Manag Executive 17(1):56–69

Hethorn J, Ulasewicz C (2008) Sustainable fashion, why now? A conversation about issues, practices, and possibilities. Fairchild Books, New York

iFixit (2015) Patagonia care & repair. https://www.ifixit.com/patagonia. Accessed Apr 2016

ING Economics Department (2015) Rethinking finance in a circular economy: financial implications of circular business models. The Netherlands

Jackson T (2009) Prosperity without growth: economics for a finite planet. Earthscan, London

LAUNCH Nordic (2016) Nordic: textiles. http://www.launch.org/challenges/nordic-textiles#readmore. Accessed Apr 2016

Lindhqvist T (1992) Extended producer responsibility as a strategy to promote cleaner production. Paper presented at the proceedings of the invitational seminar, Trolleholm Castle, Sweden

Liu CH, Chen M-C, Tu Y-H, Wang C-C (2014) Constructing a sustainable service business model: An S-D logic-based integrated product service system (IPSS). Int J Phys Distrib Logistics Manag 44(1–2):80–97

Lüdeke-Freund F (2010) Towards a conceptual framework of business models for sustainability. Paper presented at the ERSCP-EMU conference, Delft, The Netherlands

McDonough W, Braungart M (2002) Design for the triple top line: new tools for sustainable commerce. Corp Environ Strategy 9(3):251–258

Mina A, Bascavusoglu-Moreau E, Hughes A (2014) Open service innovation and the firm's search for external knowledge. Res Policy 43(5):853–866

Mont O (2008) Innovative approaches to optimising design and use of durable consumer goods. Int J Prod Dev 6(3/4):227–250

Mud Jeans (2016) http://www.mudjeans.eu/lease-philosophy/. Accessed Apr 2016

Net-works (2016) http://net-works.com/. Accessed Apr 2016

Niinimäki K, Hassi L (2011) Emerging design strategies in sustainable production and consumption of textiles and clothing. J Clean Prod 19:1876–1883

Norden (2015) Well dressed in a clean environment: nordic action plan for sustainable fashion and textiles. In N. C. o. Ministers (ed). Copenhagen

Nudie (2015) http://www.nudiejeans.com/reuse/#/nudie-jeans-good-environmental-choice/ (14 Aug 2015)

Open Garments (2009) http://www.open-garments.eu/approach.html. Accessed Apr 2016

Packard V (1963) The waste maker. Penguin, London

Pal R (2014) Sustainable business development through designing approaches for fashion value chains. In: Muthu SS (ed) Roadmap to sustainable textiles and clothing. Singapore, Springer

Pal R (2015) EPR-systems and new business models for sustained value creation: a study of second-hand clothing networks in Sweden. Paper presented at the 15th AUTEX world textile conference, Bucharest, Romania

Pal R (2016) Sustainable value generation through post-retail initiatives: an exploratory study of slow and fast fashion businesses. In: Muthu SS, Gardetti MA (eds) Green fashion, vol 1. Springer, Singapore, pp 127–158

Pedersen ER, Netter S (2015) Collaborative consumption: business model opportunities and barriers for fashion libraries. J Fashion Mark Manag 19(3):258–273

Perella M (2015) New fabrics make recycling possible, but are they suitable for high street? The Guardian. http://www.theguardian.com/sustainable-business/sustainable-fashion-blog/2015/jan/22/fabric-recycling-closed-loop-process-high-street-fashion. Accessed Apr 2016

Pure waste (2016) http://www.purewaste.org/company/about-us.html. Accessed Apr 2016

Reddy J (2014) UK fashion social enterprises support recovering addicts, refugees and ageing. The Guardian. http://www.theguardian.com/sustainable-business/sustainable-fashion-blog/uk-social-enterprise-fashion-support-refugees-women. Accssed Apr 2016

Reim W, Parida V, Örtqvist D (2015) Product–service systems (PSS) business models and tactics— a systematic literature review. J Clean Prod 97:61–75

ReLooping Fashion (2016) http://reloopingfashion.org/. Accssed Apr 2016

Re:newcell (2016) The business concept. http://renewcell.se/. Accssed Apr 2016

Retextile (2016) http://retextile.se/en/home/. Accssed Apr 2016

Rockström J, Klum M (2012) The human quest: prospering within planetary boundaries. Langenskiöld, Stockholm

Rohrbeck R, Konnertz L, Knab S (2013) Collaborative business modelling for systemic and sustainability innovations. Int J Technol Manage 63(1–2):4–23. doi:10.1504/ijtm.2013.055577

Romero D, Molina A (2011) Collaborative networked organisations and customer communities: value co-creation and co-innovation in the networking era. Prod Plan Control Manag Oper 22 (5–6):447–472

Saebi T, Foss NJ (2014) Business models for open innovation: matching heterogenous open innovation strategies with business model dimensions. Social Science Research Network (SSRN). http://papers.ssrn.com/sol3/papers.cfm?abstract_id=2493736 (22 July 2015)

Sinha P, Muthu SS, Dissanayake G (2015) Remanufactured fashion. Springer, Singapore

Smith J, Newman M (2014) Fashion can create sustainable employment for marginalised women. The Guardian. http://www.theguardian.com/sustainable-business/sustainable-fashion-blog/fashion-social-enterprise-sustainable-employment-women. Accessed Apr 2016

Stahel W (2007) Resource-miser business models. Int J Environ Technol Manage 7(5/6):483–495

Stubbs W, Cocklin C (2008) Conceptualizing a sustainability business model. Organ Environ 212:103–127

Sustainable Apparel Coalition (2016) http://apparelcoalition.org/the-coalition/. Accessed Apr 2016

Sustainable Consumption Roundtable (2006) Looking back, looking forward: lessons in choice editing for sustainability. http://www.sd-commission.org.uk/data/files/publications/Looking_back_SCR.pdf. Accessed Apr 2016

Swedish Environmental Protection Agency (2016) Environmental benefit of reuse and recycling

Teijin (2016) Closed-loop recycling system ECO CIRCLE. http://www.teijin.com/solutions/ecocircle/. Accessed Apr 2016

The North Circular (2016) http://thenorthcircular.com/about-us. Accessed Apr 2016

Tischner U, Charter M (2001) Sustainable product design. In: Tischner U, Charter M (eds) Sustainable solutions: developing products and services for the future. Greenleaf, Sheffield, pp 118–138

Tukker A (2004) Eight types of product-service system: eight ways to sustainability? Experiences from SusProNet. Bus Strategy Environ 13(4):246–260

Vincent J, Bogatyreva O, Bogatyrev N, Bowyer A, Pahl A-K (2006) Biomimetics: its practice and theory. J R Soc Interface 3(9):471–482

WBCSD (2008) Sustainable consumption facts and trends: from a business perspective. World Business Council for Sustainable Development. http://www.wbcsd.org/pages/edocument/edocumentdetails.aspx?id=142. Accessed Apr 2016

Wijkman A, Rockstrom J (2012) Bankrupting nature—denying our planetary boundaries. Routledge, New York

Worn Again (2016) http://wornagain.info/about/press-coverage/. Accessed Apr 2016

Yunus M (2007) Social business. Yunus Centre. http://www.muhammadyunus.org/index.php/social-business/social-business. Apr 2016

Yunus M, Moingeon B, Lehmann-Ortega L (2010) Building social business models: lessons from the Grameen experience. Long Range Plan 43:308–325

ZERI (2013) What is ZERI? http://zeri.org/ZERI/About_ZERI.html. Accessed Aug 2014

# Sustainability in Jute-based Industries

Sanjoy Debnath

**Abstract** Jute industry is one of the oldest textile industries next to cotton, wool and silk. A large number of people are involved directly or indirectly with this jute-based industry. Present-day sustainability of this industry is being questioned in different forums. Possible facts which are responsible to sustain this industry have been discussed elaborately. This chapter also covers different segments of this industry wherein their present scenario and future requirements for sustainability have been explained. A holistic approach has been made to cover product manufacturing, machine manufacturing, ancillary manufacturing and marketing industries associated with this industry. Diversification of process and product is an important aspect for the sustenance of this industry.

**Keywords** Diversification of jute industry · Jute-based industry · Problems in jute industry · Sustainability

## 1 Introduction: *History and Present Scenario—An Overview*

Among different ancient bast fibres, jute is second to flax origins in Mediterranean region (Lord 2003), and later, it came to India. Based on the record, jute was known as '*patta*' in 800 BC. It is popular more than a century for its industrial applications such as packaging material in different sectors, geotextile application and carpet backing (Debnath et al. 2009). Since seventeenth to twentieth century, the jute industry in India was delegated by the British East India Company, which was the first jute trader (Anonymous 2016a). Palit and Kajaria (2007) documented several historical events which were evidences for the growth of the jute industry. In 1833, the Dutch government specified bags made of jute instead of flax for carrying coffee

S. Debnath (✉)
Mechanical Processing Division, ICAR—National Institute of Research
on Jute & Allied Fibre Technology, 12, Regent Park, Kolkata 700040, India
e-mail: sanjoydebnath@yahoo.com; sanjoydebnath@hotmail.com

© Springer Nature Singapore Pte Ltd. 2017
S.S. Muthu (ed.), *Sustainability in the Textile Industry*,
Textile Science and Clothing Technology, DOI 10.1007/978-981-10-2639-3_7

from the East Indies. At that time, flax fibre was imported from Russia. But the Crimean War during 1854–56 led to the stoppage of supply of flax from Russia and this forced Dundee, the famous jute manufacturing centre of UK, to move for substitute fibre. In Dundee, then, the flax mills were converted into jute mills. On the other side, the American Civil War (1861–65) gave further impetus to the jute trade, as supplies of America cotton were much restricted (Palit and Kajaria 2007). Since then, the industry did not return to flax or cotton again. The main reason behind this shift is the cost competitiveness of the substitute fibres (jute). During twentieth century, Margaret Donnelly, a mill landowner in Dundee, first set up the jute mill in India. In the year 1793, the first consignment of Jute was exported by East India Company. Beginning of 1830, Dundee spinners have determined to spinning of jute yarn by modifying their existing power-driven flax machinery. This leads to increase in the export and production of raw jute from Indian sub-continent. In the year 1854, the first jute factory in India was established at Rishira, about 20 km north of Calcutta (Anonymous 2016b). The industry made tremendous progress in the later part of the nineteenth century. Subsequently, the industry was boosted by the two world wars. Gradually, jute became the speciality of Dundee as documented by Gray (1989) and Mokyr (2003). By the 1880s, there were over 70 jute mills and factories in Dundee. The biggest factory of all was the Camperdown works, and fourteen thousand people worked there. In 1869, five mills were established with around capacity of 950 looms. The growth was very much fast that by the year 1910, and 38 companies were operating around capacity of 30,685 looms. These rendered more than a billion yards of cloth and over 450 million bags (Anonymous 2016a). In the year 1880, jute industry has acquired almost the whole of Dundee and Calcutta. Later during the nineteenth century, the manufacturing of jute has started in other countries such as France, America, Italy, Austria, Russia, Belgium and Germany. Most of the jute barons had started to quit India, leaving the set up of jute mills after Independence, and then, Indian businessmen (mostly *Marwaris*) owned most of the jute mills. Jute industry is one of the labour-intensive industries, and hence, the jute mills slowly closed down in European countries and America due to soaring labour cost.

After agriculture, textile and garment sector is the second major job provider in India and has the ability to drive economy. But for that, India needs to explore it further to present something different to the world. Jute industry, a part of textile and garment sector, has vast hidden abilities in this regard.

In recent days, jute textile industry is one of the major industries in the eastern India, particularly in West Bengal (Anonymous 2016c). This industry supports around 40 lakh farm families and provides direct employment to 2.6 lakh industrial workers and 1.4 lakh in the tertiary sector. The production process in the jute industry passes through a variety of actions, which begins with cultivation of raw jute, processing of jute fibres, spinning, weaving, bleaching, dyeing, finishing, and marketing of both the raw jute and its finished products. These indicate clearly about the labour intensiveness of the jute industry, and as such, its labour-to-output ratio is also high in spite of various difficulties being faced by the industry. Capacity utilization of the industry is around 75 %. Jute industry contributes to the export earnings in the range of Rs. 1000–1200 crore annually.

Today, most of the jute mills in the country are of composite mills comprising spinning and weaving (power loom) sectors (Anonymous 2016d). The Indian jute industries are largely dominated and owed by private sector ownership having 93 % mills (84 out of 90 mills) owned by private companies. The Indian Government ownership is only to extent of balance 7 % through ownership by *NJMC* (). Presently, there are 90 jute mills in India, of which 67 mills are located in West Bengal, 9 mills located in Andhra Pradesh, 3 each in Uttar Pradesh, Bihar and Orissa, 2 each in Assam and Chhattisgarh and 1 in Tripura. Kaur (2014) stated that as on 1 April 2014, the jute industry provides direct employment to 215,000 workers in organized mills and indirect employment to 150,000 workers. It provides livelihood to near about four million farm families who are engaged in the cultivation of jute. As per as the production of jute is concerned, the total production of raw jute during 2013–2014 was 11,416.4 thousand bales (1 bale 180 kg) or 2.05 MT, whereas production of jute goods was 15.3 lakh MT. For the past three years, there is no major change in the production of raw material and goods. As far as the export of jute is concerned, India exported about 1.3 lakh MT worth Rs. 1593.6 crore jute and jute products during April–December 2013. The USA, the UK, Saudi Arabia, Germany, Egypt and Turkey are the major importing countries. Table 1 presented the total installed capacity of the jute industry.

It is clear from Table 1 that 75 % of the jute goods produced are used as packaging materials, hessian and sack fabrics. The other products are carpet yarn, cordage, felts, padding, twine, ropes, decorative fabrics, geotextile fabrics and miscellaneous items for industrial uses. The productions (in 000 MT) of hessian fabric, sacking fabric, carpet backing cloth (CBC) and other jute goods during the last five years are shown in Table 2.

It is clear from the table that the overall production of hessian, sacking, CBC and yarn consumption/demand shows decreasing trend, and for other jute products, the trend remains unchanged. Many of the jute industries still use jute batching oil which contains some harmful chemical compounds. The export market today is

**Table 1** Aggregated installed capacity in jute industry (Anonymous 2016d)

| Processing stage | Product type | Spindles (numbers) | Looms (numbers) | Total installed capacity (000 TPA) |
|---|---|---|---|---|
| Spinning | Fine yarn | 630,776 | – | 2732 |
| | Coarse yarn | 132,904 | – | |
| Weaving | Hessian | – | 18,637 | |
| | Sacking | – | 28,592 | |
| | CBC* | – | 905 | |
| | Others | – | 3223 | |
| Total | | 763,680 | 51,357 | |

*Carpet backing cloth

**Table 2** Production of jute goods in India (Anonymous 2016d)

| Period (April–March) | Hessian | Sacking | CBC | Yarn | Others | Total |
|---|---|---|---|---|---|---|
| 2010–11 | 244.4 | 1076.9 | 4.7 | 177.3 | 62.4 | 1565.7 |
| 2011–12 | 239.9 | 1165.1 | 3.9 | 123.4 | 50.4 | 1582.4 |
| 2012–13 | 210.0 | 1218.2 | 2.7 | 114.1 | 46.3 | 1591.3 |
| 2013–14 | 202.5 | 1150.4 | 3.2 | 109.0 | 62.6 | 1527.4 |
| 2014–15 | 211.3 | 901.8 | 2.7 | 90.6 | 60.8 | 1267.2 |

**Table 3** Export trend of jute goods from India (Anonymous 2016d)

| Product | 2010–11 | 2011–12 | 2012–13 | 2013–14 | 2014–15 |
|---|---|---|---|---|---|
| Hessian | 7405 | 9788 | 9033 | 8610 | 7696 |
| Sacking | 2239 | 4189 | 4165 | 5270 | 2966 |
| Yarn | 5310 | 2820 | 2212 | 1436 | 1387 |
| JDPs | 2697 | 3199 | 3636 | 4839 | 5086 |
| Others | 890 | 954 | 872 | 1064 | 1003 |
| Total | 18,541 | 20,950 | 19,918 | 21,219 | 18,138 |

emphasizing more on the eco-friendliness process. This may be one of the probable reasons to show decreasing trend. Another possible reason is the stiff competition with the similar synthetic counterpart whose market strongly dominated by the synthetic industrial lobby apart from the cost competitiveness while compared to jute. Further, if we move in detailed search, this study also shows the export trend of jute goods in India (Table 3).

It is depicted from Table 3 that the hessian fabric export trend fluctuates and the trend of jute sacking and yarn reduces drastically. However, the non-traditional products such as jute non-woven, geotextile, agrotextile and insulation material which come under other categories, show an increasing trend. The jute being the natural material has more acceptability in ecosystem, and hence in the recent times, the demand of such environmentally friendly product demand increases. The study (Anonymous 2016d) shows that there are more export demands in the area of JDPs (jute diversified products) in India. India being the major producer of jute fibre, the many self-help groups (SHGs) and NGO (non-governmental organization) are being started to produce the jute diversified products and export them in European market. This generates more revenue generation, and with time, the companies involving with this business are progressing well. Further, to establish these facts, Table 4 presents the exports (value in INR Million) of JDPs from India.

It is clear from the trend of Table 4 that the different JDPs (floor coverings, carry bags and other JDPs) increase which show that the future sustenance of this jute-based industry lies on the diversification through various JDPs. Since the raw material used is the natural product, the final product is also natural, provided the industry must go for substitute of JBO (jute batching oil) which is extracted from the petroleum-based industries. Today some of the mills have adopted RBO

**Table 4** Export trend of JDPs from India (Anonymous 2016d)

| Products | 2010–11 | 2011–12 | 2012–13 | 2013–14 | 2014–15 |
|---|---|---|---|---|---|
| Floor coverings | 1342 | 1420 | 1790 | 2161 | 2389 |
| Carry bags | 1264 | 1637 | 1697 | 2448 | 2518 |
| Other JDPs | 91 | 142 | 158 | 230 | 179 |
| Total | 2697 | 3199 | 3636 | 4839 | 5086 |

(rice bran oil), a plant extract used as substitute of JBO to make the process an well as the product green and sustainable.


## 2 Problems in Jute-based Industries

In India, this industry suffered a serious setback in 1947 due to the partition of the subcontinent (Anonymous 2016b). After partition, about 80 % of the jute-growing areas went to East Pakistan (Bangladesh), while nearly 90 % jute mills remained in India. In 1959, the international demand of jute products decreased substantially as a result of which 112 jute factories were closed down. At present, there are only 60 jute-producing mills in India. Most of these mills are along the Hooghly River, especially to the north of Kolkata.

Most of the jute-based industries still today are being producing the age-old products such as jute sacking and hessian as packaging material and to some extent carpet backing. These products in total account around 95 % of the total production of the industry. Only countable industries are involved in diversified product development process for commercial purposes. These products are mostly laminated jute fabric, geotextile, industrial textiles, etc. In present days, lot of awareness have been taken place in the area of eco-friendliness. In this regard, the traditionally practiced jute batching oil used during spinning is found to have some carcinogenic components which may be contaminated during packaging as food grain. Efforts have been made to replace this jute batching oil, but till date, no such substitute has been developed. Researchers have tried with eco-friendly vegetable oil (rice bran oil (RBO), linseed oil, jute seed oil, castor oil, etc.) based emulsion for jute spinning, but none of them are found similar/better spinning performance in all respect compared to traditional jute batching oil (Basu et al. 2009). In spite of different spinning problems with other than jute batching oil, some of the jute industries are using RBO in export products especially in the area of food grain packaging to eliminate the carcinogenic component contamination during food grain packaging.

According to the Jute Packaging Norms and Legal Protection to Jute Cultivators the Parliament of India had enacted the JPM (Jute Packaging Mandatory) Act 1987 with an objective to protect the jute industry (Anonymous 2016c). As per this act, the food grain and sugar produced are reserved and mandatorily packed in jute bags manufactured every year. The Government of India recently found that the jute industry could not match demands in 2011–2012 for supply of 13 lakh bales or 4.33

lakh tons of gunny bags for Rabi crop supply of 2012–2013. Government said that with 10 mills remaining closed, the jute industry is short in capacity by 1.5 lakh ton. Presently, it can produce 11 lakh tons of jute sacks/gunny bags. Its installed capacity, however, is 15.02 lakh tons, and assuming a 83 % utilization, its stated capacity is 12.47 lakh tons. The industry earns a business of around Rs. 10,000 crore by selling its entire produce to FCI (Food Corporation of India), sugar mills, and co-operatives in the Indian market apart from the export. FCI makes a bulk purchase of almost 35–40 % of jute mills produce. In 2012–2013, FCI is expected to purchase 6.34 lakh tons and 4.33 lakh jute/gunny bags (Anonymous 2016c).

Apart from the above problems, in India, jute industry suffers lot from different political interference, labour problem, jute mills owner are mostly headed by business community rather than entrepreneurs, and shortage of jute fibre supply due to low rainfall, which also lead against the sustainability of the jute industry (Kundu et al. 1959).

## 3   Sustainability: Diversification of Product and Process

Jute is also known as the 'golden fibre', a plant that produces a fibre mainly used for sacking and cordage and carpet backing (Anonymous 2016b). This raw material is used for sacks globally which is one of the most versatile fibres of nature. Still jute is cheaper and plays an important among all textile fibres next to cotton. The jute mills are integrated units consisting of both spinning and weaving units. The main products of jute industry are gunny bags, canvas, pack sheets, cotton–jute, paper-lined hessians, hessian cloth, carpets, carpet backings, rugs, cordage and twines.

The Indian jute industry has been expanding really fast, spanning from a wide range of lifestyle consumer products, with courtesy to the versatility of jute. By the innovative ways of bleaching, dyeing, and finished goods processes, the jute industry now provides finished jute products that are softer and have lustre with aesthetic appeal. Today, jute has been defined as eco-friendly natural fibre with utmost versatility ranging from low-value geotextiles to high-value carpets, decorative, apparels (Debnath et al. 2007a, b), composites, upholstery furnishings, etc. (Debnath et al. 2009). In the same line of development, Sengupta and Debnath (2010, 2012) documented jute-based products for upholstery application. They also compared their developed jute-based products with commercial non-jute similar products. Debnath et al. (2007a, b) developed jute and hollow-polyester-blended bulked yarn for warm fabrics such as knitted sweater and jacket, and they found that the bulkiness of the jute-polyester-bulked yarn is superior to jute yarn.

There are several research institutes whose researches are mostly concentrated on agricultural development of jute fibre in India (ICAR Central Research Institute of Research on Jute & Allied Fibres, Barrackpore, West Bengal) and Bangladesh (Bangladesh Jute Research Institute, Dhaka). These researches lead to good varietal development for fine yarn spinning, improvement in yield and productivity of jute, and improved retting process and technology. Globally, it has been accepted that good and fine jute fibre is essential for the development of diversified value-added products. One can look into the important properties of jute fibre since it has huge diversifying potentiality. Advantages of jute include good insulating and antistatic properties, as well as having low thermal conductivity and a moderate moisture regain (Anonymous 2016e). It includes acoustic insulating properties and manufacture with no skin irritations. Jute has the ability to be blended with other fibres, both synthetic and natural, and accepts cellulosic dye classes such as natural, basic, vat, sulphur, reactive, and pigment dyes. While relatively cheap synthetic materials in many uses are replacing jute, jute's biodegradable nature is suitable for the storage of food materials, where synthetics would be unsuitable.

## 4   Sustainability and Sustenance of Jute-based Industries

The jute industry occupies significant place in the Indian economy (Anonymous 2016a). The Indian jute industry is a very old and predominant in the eastern part of India. The Government of India has included the jute industry for special attention in its National Common Minimum Programme. It forms an integral part of the Indian textile industry. Further, the jute industry contributes to the national exchequer from exports and taxes (Bhattacharya 2013).

As per the sustenance and sustainability of jute industry are concerned, there come some of the following important aspects: there are lot of changes required to be implemented in the present jute industries, and this may certainly help in diversification. Majority of the industries are using the age-old machinery. Even the new jute industries installed in Bangladesh and India are also using the used jute spinning machinery either from India or China. The old machinery consumes more energy for the production of same quantity of yarn. The old machinery is also having lesser productivity and requires frequent maintenance. This caused increase in production cost or, in other term, increase the conversion cost from fibre to yarn. Almost all the jute industries are still concentrating on the traditional products such as jute sacks and hessian. In the era of synthetics, similar synthetic products are offering in the market of at least few folds lesser in price. This creates stiffer competition in the existing market. The probable solution is to move towards diversification from the conventional products. It has been proven that through diversification, there is always higher cost-to-benefit ratio. Day by day, the cost of fibre production (cultivation and fibre extraction) becomes increasing, as a result the jute mill owners find difficult to push their products in the competitive market. Unlike other agricultural produces, the cultivable land is reducing due to the expansion of the urbanization. On the other hand, the farming community is also shifting from jute to other crops due to uncertainty of the rain. The jute retting process requires a huge quantity of water. Due to the climate change and change in the rainfall pattern, the farmers are suffering from scarcity of water during the retting season of jute fibre. As a result, the cost of fibre is increasing every day.

On the contrary, due to advancement of science and technology, the synthetic products are becoming cheaper at the cost of environmental. Many of the jute industries are being closed few months of a year due to shortage of raw material (jute fibre). Even the quality of the fibre produced in low water retting decreases the quality of the retted jute fibre. On the one hand, industry finds stiff competition with similar synthetic products in the market with lower cost, and on the other hand, the existing jute fibre is not sufficiently in good quality.

Only countable industries are involved in diversified product development process for commercial purposes. These products are mostly polyethylene-/polypropylene-laminated jute fabric (raw hessian fabric, bleached fabric, printed fabric, dyes fabric, etc.). Compared to these products, much lesser quantity of jute-cotton-blended fine fabric dyed, designed, laminated or non-laminated fabrics are being produced. These decorative fabrics are mostly used for diversified fashion bag and utility bags (water bottle bag, win bottle bag, tea coaster, shopping bag, school bag, backpack, soft luggage bag, etc.), material for home decoration (window/door curtain, table cover, lamp shed, furnishing material, etc.), car inter lining (seat cover, inner roof top, foot mat/carpet, inner lining of door, etc.) and other fashion products. Jute non-woven and woven materials are used for the development of polyester resin fibre-reinforced composite material for different industrial applications in automobile industry as car door panel, dash board, etc.

## 5 Conclusions and Impending

Application of jute area must be increased. India needs to work on quality by adopting new technologies (Kaur 2014). Jute Research Association, such as ICAR–NIRJAFT, Kolkata, IJIRA, Kolkata, Department of Jute and Fibre Technology, Kolkata, must work together to utilize resources for the betterment of the industry. Government must make efforts in R&D to strengthen the crisis-stricken but one of the oldest industries in India that is the jute industry.

Furthermore, jute industry should concentrate more towards diversification from its existing products. These will fetch more profit and has less market competitor (synthetic counterpart). The jute mill owners should also need to change their mindset to overcome the present situation and move towards the sustenance of this traditional industry. Still today, a majority portion of people is involved directly/indirectly with this jute industry wherever it exists. Last but not the least, the modernization in jute industry is foremost part as per as the sustainability of this jute industry is concerned. Lastly, the application of plant based substitute jute conditioning oil in place of JBO will lead towards greener process and product. Overall, the JDPs and geotextile made from these eco-friendly substitute of JBO may sustain the jute industry rather than the traditional sacks and hessian fabrics.

# References

Anonymous (2016a) http://www.indianmirror.com/indian-industries/jute.html. Dated 03 Feb 2016

Anonymous (2016b) http://www.vajiramandravi.in/jute-textile-in-india.html. Dated 03 Feb 2016

Anonymous (2016c) http://www.gktoday.in/jute-industry-of-india/. Dated 03 Feb 2016

Anonymous (2016d) Sonar Beni-the golden fibre: success stories. National Jute Board, Kolkata, pp 12–17

Anonymous (2016e) http://www.worldjute.com/about_jute/abj_intro.html, Dated 03 Feb 2016

Basu G, De SS, Samanta AK (2009) Effect of bio-friendly conditioning agents on jute fibre spinning. Ind Crops Prod 2(9):281–288

Bhattacharya B (2013) Marketing of raw jute. In: Advances in jute agronomy, processing and marketing of raw jute. PHI Learning Limited, New Delhi, pp 144–154, ISBN: 978-81-203-4670-3

Debnath S, Sengupta S, Singh US (2007a) Properties of jute and hollow-polyester blended bulked yarn. Journal of The Institution of Engineers (India). Text Eng 87(2):11–15

Debnath S, Sengupta S, Singh US (2007b) Comparative study on the physical properties of jute, jute-viscose and jute-polyester (hollow) blended yarns. Journal of The Institution of Engineers (India). Text Eng 88(1):5–9

Debnath S, Roy AN, Basu G, Chattopadhyay SN (2009) NIRJAFT's technologies for rural development. In: Shukla JP (ed) New technologies for rural development having potential of commercialisation. Allied Publishers Private Limited, Delhi, pp 136–142, ISBN: 978-81-8424-442-7

Gray A (1989) A history of Scotland: Modern times. Book five. Oxford University Press, Oxford, New York, pp 3–4

Kaur R (2014) http://www.mapsofindia.com/my-india/india/one-of-the-oldest-but-crisis-stricken-jute-industry-in-india. Dated 03 Feb 2016

Kundu BC, Basak KC, Sarcar PB (1959) Jute in India. The Indian Central Jute Committee, Calcutta, pp 1–395

Lord PR (2003) Handbook of yarn production: Technology, Science ane Economics. Woodhead Publishing Limited, Cambridge, England, p 2, ISBN: 9781855736962

Mokyr J (2003) The Oxford encyclopedia of economic history. Oxford University Press, Oxford, New York, p 213

Palit S, Kajaria S (2007) Jute industry—a historical perspective. Firma KLM Private Limited, Kolkata, pp 1–546

Sengupta S, Debnath S (2010) A new approach for jute industry to produce fancy blended yarn for upholstery. J Sci Ind Res 69(12):961–965

Sengupta S, Debnath S (2012) Studies on jute-based ternary blended yarn. Indian J Fibre Text Res 37(3):217–223

Printed in the United States
By Bookmasters